# Communications in Computer and Information Science 1457

T0171825

More information about this series at https://link.springer.com/bookseries/7899

George Fletcher · Keisuke Nakano ·
Yuya Sasaki (Eds.)

# Software Foundations
# for Data Interoperability

5th International Workshop, SFDI 2021
Copenhagen, Denmark, August 16, 2021
Revised Selected Papers

 Springer

*Editors*
George Fletcher 🆔
Eindhoven University of Technology
Eindhoven, The Netherlands

Keisuke Nakano 🆔
Tohoku University
Sendai, Japan

Yuya Sasaki 🆔
Osaka University
Osaka, Japan

ISSN 1865-0929 ISSN 1865-0937 (electronic)
Communications in Computer and Information Science
ISBN 978-3-030-93848-2 ISBN 978-3-030-93849-9 (eBook)
https://doi.org/10.1007/978-3-030-93849-9

This Springer imprint is published by the registered company Springer Nature Switzerland AG
The registered company address is: Gewerbestrasse 11, 6330 Cham, Switzerland

# Preface

This volume contains the papers presented at Fifth Workshop on Software Foundations for Data Interoperability (SFDI 2021) held on August 16, 2021, as a hybrid conference: physically in Copenhagen, Denmark, for all attendees who could travel to Copenhagen and remotely for everybody else.

The SFDI workshop aims at fostering discussion, exchange, and innovation in research and development in software foundations for data interoperability as well as the applications in real-world systems such as data markets. We invite not only people from the database community but also researchers and professionals from the areas of programming language, software engineering, distributed computing, artificial intelligence, and natural language processing to share their knowledge and experience.

For example, bidirectional transformation has been intensively studied in both database and programming language communities to address the view update problem, and a recent trend is to use bidirectional transformation for data sharing and data interoperability in P2P networks. In addition, many machine learning and natural language processing models have been utilized for data management, especially for data integration tasks. We thus hope to bring together academia and industry people in related areas for the discussion of challenges and technical solutions to software foundations of data interoperability.

SFDI 2021 adopted a single-blinded peer review process. Each submission was reviewed based on relevance, novelty, technical depth, experiments, presentation, etc., by three Program Committee (PC) members and external reviewers. After careful and thorough discussions, the PC accepted five papers from eight submissions. The program also included two invited talks by Juan Sequeda and Perdita Stevens. This post-proceedings volume contains an extended abstract of Perdita's talk and a summary of the BISCUITS project led by Zhenjiang Hu, Masatoshi Yoshikawa, and Makoto Onizuka.

We would like to thank all invited speakers and authors for their contributions. We are grateful to all PC members for their hard work and the help of the EasyChair conference management system for making our work of organizing SFDI 2021 much easier. This workshop was supported by JSPS KAKENHI Grant Numbers JP18H04093 and JP17H06099.

November 2021

George H. L. Fletcher
Keisuke Nakano
Yuya Sasaki

# Organization

## Program Chairs

| | |
|---|---|
| George H. L. Fletcher | Eindhoven University of Technology, The Netherlands |
| Keisuke Nakano | Tohoku University, Japan |
| Yuya Sasaki | Osaka University, Japan |

## Program Committee

| | |
|---|---|
| Muhammad Aamir Cheema | Monash University, Australia |
| Renzo Angles | Universidad de Talca, Chile |
| Yang Cao | Kyoto University, Japan |
| Yuyang Dong | NEC Corporation, Japan |
| Stefania Dumbrava | ENSIIE Paris-Evry, France |
| Claudio Gutierrez | Universidad de Chile, Chile |
| Ekaterini Ioannou | University of Tilburg, The Netherlands |
| Verena Kantere | University of Ottawa, Canada |
| Sebastian Maneth | Universität Bremen, Germany |
| Victor Marsault | Université Paris-Est, France |
| George Papadakis | National Technical University of Athens, Greece |
| Jianbin Qin | Shenzhen University, China |
| Oscar Romero | Universitat Politècnica de Catalunya, Spain |
| Toshiyuki Shimizu | Kyoto University, Japan |
| Dimitrios Skoutas | Athena Research Center, Greece |
| Massimo Tisi | IMT Atlantique, France |
| Yannis Velegrakis | Utrecht University, The Netherlands |
| Hannes Voigt | Neo4j, Germany |
| Jiannan Wang | Simon Fraser University, Canada |
| Xiang Zhao | National University of Defense Technology, China |
| Erkang Zhu | Microsoft, USA |

## Steering Committee

| | |
|---|---|
| Makoto Onizuka | Osaka University, Japan |
| Masatoshi Yoshikawa | Kyoto University, Japan |
| Zhenjiang Hu | Peking University, China/NII, Japan |

## Additional Reviewers

Kota Miyake
Vandang Tran

# Contents

# Invited Papers

# Theory and Practice of Networks of Models

Perdita Stevens(✉) ⓘ

Laboratory for Foundations of Computer Science School of Informatics
University of Edinburgh, Edinburgh, UK
Perdita.Stevens@ed.ac.uk
http://homepages.inf.ed.ac.uk/perdita

**Abstract.** Separating concerns into multiple data sources, such as multiple models of a software system under development, enables people to work in parallel. However, concerns must also be re-integrated, and this gives rise to many interesting problems, both theoretical and practical. In my keynote talk i will discuss some of them; in this accompanying paper i give some background to my work in recent years and summarise some key points and definitions.

**Keywords:** Bidirectional transformation · Consistency maintenance · Megamodel

## 1 Introduction

The power of information technology to benefit our lives comes from bringing together vast amounts of data, and complex behaviour based on that data, which unaided human brains could not handle. This, however, makes the development and maintenance of the information technology difficult for us: humans have to be able to direct it, without any one of us fully understanding it.

Separation of concerns, in Dijkstra's famous terminology [2], is fundamentally the only hammer we have to tackle this problem. Because no individual human can hold in mind *all* the information that is relevant to even a moderately-sized information system, it is essential for us to be able to parcel out that information so that the part of it that is necessary to each decision can, indeed, be held in mind and manipulated as necessary.

A related, but distinct, problem is the need to hold the data we care about in different places – on different computers, sometimes in different geographical locations, sometimes in different legal jurisdictions or on opposite sides of "Chinese walls". Sometimes it is impractical to colocate the data; sometimes it is positively undesirable, illegal or unethical [4].

We would not be considering, together, information that is held separately, if there were no circumstance under which it is somehow related. Perhaps some computation is to be done that draws on data from two separate sources, thereby producing a third set of data. Whenever such a situation pertains, there is the

G. Fletcher et al. (Eds.): SFDI 2021, CCIS 1457, pp. 3–12, 2022.
https://doi.org/10.1007/978-3-030-93849-9_1

possibility of getting a wrong, or nonsensical, answer – that is, it is possible for there to be inconsistencies between the data sources, and/or errors in the relationships that are assumed or imposed on them. The imposition and maintenance of such relationships is called the *consistency maintenance problem*. It may arise in very simple guises: perhaps a datum is intended to have duplicate copies in two locations, so that if the copies differ at all, the sources are inconsistent. Or it may involve arbitrarily complex relationships.

In the simplest of cases, the dependencies between data sources flow in one direction only, producing an acyclic graph. For example, if one or more independent data sources are regarded as authoritative, and another is to be computed from them, then the only kind of inconsistency that can arise is if the computed data becomes out-of-date with respect to the sources. In such a case consistency can be restored by recomputing the outdated data. There may still be challenging problems of how to do so efficiently, but we do not encounter *bidirectionality*.

The essence of bidirectionality is given [6] by the following three features:

1. There is separation of concerns into explicit parts such that
2. more than one part is "live", that is, liable to have decisions deliberately encoded in it in the future; and
3. these[1] parts are not orthogonal. That is, a change in one part may necessitate a change in another.

The first point gives us data in separate places; the third tells us that inconsistency is a possibility; the second rules out mere recomputation as a solution.

Bidirectionality can be, and often is, attacked as a purely theoretical topic in which the first thing we do is to assume that there is a single consistent metadata framework into which all the data of interest fits; then the problem of maintaining consistency can be attacked within that framework. Unfortunately the real world is not so neat. We need to contend with

- data governance issues such as the need to control which data sources may be modified when – because our consistency maintenance framework may not always have authority to modify data in arbitrary ways
- legacy data sources – because we may not be allowed to reengineer a data source to conform to our chosen framework
- legacy programmed transformations or other consistency maintenance mechanisms between data sources – because it may not be desirable or practicable to recreate them
- the need to support fundamentally different notions of what it is for (even the same) data sources to be consistent – because we may need more or less stringent notions of consistency at different times, as shared meanings converge or diverge

---

[1] The original version says "the": but the key point is that live parts, which therefore cannot be simply overwritten by recomputation, should have dependencies between them.

– the need to support gradual adoption of automated consistency checking and maintenance – because in non-toy problems, big bang adoption of automated consistency maintenance will seldom be practical even if it is desirable.

There is a saying that there is no problem in software engineering that cannot be solved by adding another level of indirection[2]. We shall see that this may be a case in point – although many problems remain to be solved.

## 2   Bidirectional Transformations

*Bidirectional transformation* (bx, for short) is the term we use for any automated way to check and restore consistency between two or more data sources. A bidirectional transformation might be written in a specialist bidirectional transformation language, or in a conventional (unidirectional) programming language; in the latter case it may in fact be a collection of programs, each doing part of the job. For example, there might be one Java program that looks at two sources and returns true or false depending on whether they are currently consistent, while another Java program take the same two sources and return a modified version of the first, modified in such a way as to bring it into consistency with the second, and yet a third changes the second source instead.

Most work on bidirectional transformations has involved just two data sources, for example, a source (containing all the information we know about) and a view (containing only a subset of the information). (This situation is the familiar *view-update problem* from databases.) Even with this restriction, there are many choices of setting to be made. Should the bx operate purely on the data sources themselves? Should it have access to additional information, such as intensional edits that give clues as to what a user who changed one source was trying to achieve? These can be very helpful when trying to make "corresponding" changes to the other data source. Or should it maintain and use trace information (such as the correspondence graph used by triple graph grammars), specifying at a fine-grained level which parts of one data source correspond to which parts of another? How should the decision be made about when to invoke the transformation? Under what circumstances should the transformation be allowed to fail, and what should happen if it does – e.g. should it then make no changes at all to the data sources, or should it improve consistency even if it cannot perfectly restore it? Etc. In this work we use the simplest possible formalisation, in which only the data sources themselves are available to the consistency restoration process. This has the practical advantage that it does not rely on changes to the data sources being made with a consistency-restoration-aware tool.

Up to now, mindful of audience, this paper has used the term "data source" – from now on we shall use the term "model", in accordance with the literature we are discussing, which is usually motivated by the needs of software development in general and model-driven engineering in particular. However, this is a

---

[2] See https://en.wikipedia.org/wiki/Fundamental_theorem_of_software_engineering.

distinction without a difference, since our notion of model is so general that it encompasses any data source. A point to note is that, as is conventional in the field, the term "model" sometimes means a specific collection of data with the values it takes at one instant, and sometimes means a conceptual grouping of data, encompassing all the values that it could take. We use capitals $M$ etc. for the latter notion, which we also sometimes refer to as "model set" or "model space" when plain "model" seems likely to cause confusion. We use lowercase $m$ etc. for the former and sometimes write "model instance" or "state of a model" to disambiguate.

## 3   Networks of Models

In practice two models are frequently not enough: we must separate our information into more than two concerns. This raises several new problems.

The first is how to conceptualise consistency between our $n$ concerns. There are basically two things we can do.

1. We can think of consistency as an $n$-ary relation, constraining all of the models simultaneously (for example, by specifying consistency using a set of constraints in which any constraint may mention any of the models). We can reify the consistency relation – the set of consistent tuples of states of all the models – as a model in its own right, which then conceptually contains *all* of the information, and becomes a central model, of which each of the original models is a view (Fig. 1).
2. We can embrace a network view, in which some collections (perhaps only pairs) of the $n$ models have separately-specified consistency relations placed on them. The whole network is consistent iff all of these consistency relations hold (Fig. 2).

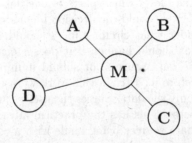

**Fig. 1.** Network with a central all-containing model: each $m \in M$ represents a globally consistent tuple of models $(a, b, c, d)$ with $a \in A$ etc., and the derived consistency relation between $A$ and $M$ is that $a'$ is consistent with $(a, b, c, d)$ iff $a' = a$, etc.

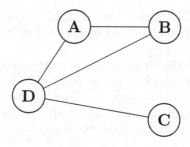

**Fig. 2.** Network without an added central model: consistency is represented directly in the relations between pairs of models.

The second view can of course be specialised to the first, provided there is no limit on the arity of the consistency relations used in the network. It becomes clearly different, however, if we constrain its relations to be binary, relating just two models each, so that the network is a graph (rather than a hypergraph).

Although the first view has an appealing simplicity, and may be fruitful when conditions permit, the second has practical advantages, as explored in [7,8]. Although there exist $n$-ary relations that cannot be expressed as a conjunction of binary relations on pairs of their $n$ places, these are cognitively difficult for humans to handle. Further, the need to handle a heterogeneous collection of models and transformations including legacy components strongly suggests that it is better to be able to work with a network of models.

Our first thought may be, then, to work with a network of models connected by binary bidirectional transformations. We may consider how to use sequences of the consistency restoration facilities of these bx to restore consistency in the network as a whole. Unsurprisingly, but unfortunately, naive approaches quickly run into problems. Given a fixed set of bx and model spaces, here are three problems that may easily arise.

1. There may be no completely consistent state of the network, e.g. because different consistency relations impose incompatible demands on the same model.
2. There may be such a state, but there may be no way to reach it using the provided bx in any sequence (see e.g. Example 5 of [7])
3. There may be many different such states, so that the overall state of the network which is reached after consistency restoration is sensitive to the precise sequence of consistency restorations chosen.

In [8] we start to explore the possibility of tackling these and other problems by adding another level of indirection (*builders*) to permit heterogeneous model and bx technologies to coexist, while simultaneously using an *orientation model* to tackle the data governance and other issues that make some sequences of consistency restoration functions acceptable and others not. In brief an orientation model records what models and relationships between them are present, and specifies which models may be modified by the consistency restoration process and which, by contrast, are currently to be considered *authoritative* and left alone; it may also specify that certain bx are to be applied only in a certain direction, thereby controlling which changes will "win" in certain conflict situations.

Taking advantage of the similarity between the problem of restoring consistency in a network of models, and the problem of incrementally building a software system from a set of sources, we adapt the pluto build system of Erdweg et al. [3], together with its underlying theory and proofs of correctness and optimality, to this setting. In very brief, the key points are as follows.

- Each model which we may ever want updated by the consistency restoration process is equipped with a *builder* whose job is, on command, to manage the process of bringing this model into consistency with a given set of its neighbours.
- The builder might be very simple, e.g. it might simply invoke one or more legacy bx in a fixed sequence; or it might be arbitrarily complex, involving some legacy bx, some explicit modifications of the model, user interaction, search...
- What the builder has to guarantee is (a) that if it terminates successfully, then it has indeed brought its model into consistency with the specified neighbours (b) that it obeys certain rules about how it must record what information it uses, and which other builders it causes to be invoked.
- Given a set of builders that provides those guarantees, the MegaModelBuild adaptation of the pluto build system can be requested to build a particular model: it will invoke the correct builders in the correct order to guarantee appropriate correctness and optimality conditions hold.

Part of the information recorded by each builder pertains to which aspects of a model are important to its consistency with its neighbours. These *stamps* can be seen as specifying what changes to a model definitely do *not* break consistency. We turn next to considering various equivalences on models that support the restoration of consistency in networks of models.

## 4    Dissecting Models and Transformations

Several different notions of sub-models appear, when we need to reason about consistency in networks of models. Let us lay three of them on the table.

We need notation for a (set-based relational) bx: a bidirectional transformation $R : M \rightleftharpoons N$ between non-empty sets of models $M$ and $N$ is given by specifying a *consistency relation*, also by slight abuse of notation called $R \subseteq M \times N$, together a pair of functions

$$\overrightarrow{R} : M \times N \rightarrow N$$

$$\overleftarrow{R} : M \times N \rightarrow M$$

whose task is to enforce consistency. So, for example, if the current models (model instances, or states of the models) are $m \in M$ and $n \in N$, we say $R(m, n)$ holds iff $m$ and $n$ are to be deemed consistent according to the bx $R$. Given current models $m$ and $n$, $\overrightarrow{R}(m, n)$ represents a new model from $N$, to be thought of as a version of $n$ that has been updated in order to bring it into consistency with $m$.

We will assume that all bx mentioned are *correct* (consistency restoration really does restore consistency, e.g. $R(m, \overrightarrow{R}(m, n))$) and *hippocratic* (consistency restoration does not alter models that are already consistent, e.g. $R(m, n) \Rightarrow \overrightarrow{R}(m, n) = n$). We will sometimes restrict attention to bx that are *history-ignorant* (restoring consistency with a second model completely overwrites the effect of restoring consistency with a previous model, e.g. $\overrightarrow{R}(m', \overrightarrow{R}(m, n)) = \overrightarrow{R}(m', n)$).

## 4.1  Coordinate Grid and History-Ignorance

**Definition 1.** *(from [5][3]) The equivalence relations $\sim_{RF}^{M}$ and $\sim_{RB}^{M}$ on $M$, and $\sim_{RF}^{N}$ and $\sim_{RB}^{N}$ on $N$, are defined as follows:*

- $m \sim_{RF}^{M} m' \Leftrightarrow \forall n \in N. \overrightarrow{R}(m, n) = \overrightarrow{R}(m', n)$
- $m \sim_{RB}^{M} m' \Leftrightarrow \forall n \in N. \overleftarrow{R}(m, n) = \overleftarrow{R}(m', n)$

*and dually,*

- $n \sim_{RF}^{N} n' \Leftrightarrow \forall m \in M. \overrightarrow{R}(m, n) = \overrightarrow{R}(m, n')$
- $n \sim_{RB}^{N} n' \Leftrightarrow \forall m \in M. \overleftarrow{R}(m, n) = \overleftarrow{R}(m, n')$

It turns out (see [5]) that knowing the $\sim_{RF}^{M}$ and $\sim_{RB}^{M}$ class of a model $m$ uniquely identifies it, and that in a precise sense these aspects of $m$ separate the information which is relevant to $N$ from that which is not. An important special case arises when these two kinds of information are sufficiently independent of one another. In that case the bx is history-ignorant and is *full* with respect to these equivalences; any choice of $\sim_{RF}^{M}$ and $\sim_{RB}^{M}$ class corresponds to a (necessarily unique) model.

## 4.2  Parts and Non-interference

**Definition 2.** *(from [7]) Let $C$ be a set of models. A part relative to $C$ is a set $P$ together with a surjective function $p : C \rightarrow P$. The value of that part in a particular model $c \in C$ is $p(c)$.*

This simple idea is related to, but more general than, the idea of a view: it may make sense to talk about parts, e.g. to talk about two model instances having the same value of a given part, even if it does not make sense to expect

---

[3] Notation slightly adapted since we here want to refer to equivalences for multiple bx: in subscripts like $RF$ and $RB$, $R$ specifies the bx while $F$, $B$ stand for forward, backward respectively.

to update a part and have some notion of the right way to update the rest of the model.

Of course, given any bx $R$ relating $C$ to some other model, the set of $\sim_{RF}^{C}$ (rsp. $\sim_{RB}^{C}$) classes together with the quotient map onto it is an example of a part relative to $C$. To say that two models have the same value of the $\sim_{RF}^{C}$ part is just another way of saying that they are $\sim_{RF}^{C}$ equivalent.

It turns out (see [7]) that we can use this idea to formalise the notion of different bx that both involve the same model caring about different parts of their common target, which is important when we want to consider whether it matters in which order we apply the bx.

**Definition 3.** *Consistency restorers* $\overrightarrow{R} : A \times C \to C$ *and* $\overrightarrow{S} : B \times C \to C$ *are* non-interfering *if for all* $a \in A, b \in B, c \in C$ *we have*

$$\overrightarrow{S}(b, \overrightarrow{R}(a,c)) = \overrightarrow{R}(a, \overrightarrow{S}(b,c))$$

**Definition 4.** *Given correct and hippocratic* $R : A \gtrless C$ *and* $S : B \gtrless C$, *an* $A/B/rest$ *decomposition of* $C$ *is a triple of parts:*

- $f_A : C \to C_A$
- $f_B : C \to C_B$
- $f_{rest} : C \to C_{rest}$

*such that the parts determine the whole, that is, if* $f_A(c_1) = f_A(c_2)$ *and* $f_B(c_1) = f_B(c_2)$ *and* $f_{rest}(c_1) = f_{rest}(c_2)$, *then* $c_1 = c_2$.

**Definition 5.** *Suppose we have correct and hippocratic* $R : A \gtrless C$ *and* $S : B \gtrless C$. *An* $A/B/rest$ *decomposition* $(f_A, f_B, f_{rest})$ *of* $C$ *is* non-interfering *if:*

1. $\overrightarrow{R}$ *only ever modifies the* $f_A$ *part: that is, for all* $a \in A$ *and* $c \in C$ *we have*

$$f_B(c) = f_B(\overrightarrow{R}(a,c))$$

$$f_{rest}(c) = f_{rest}(\overrightarrow{R}(a,c))$$

   *and dually,* $\overrightarrow{S}$ *only ever modifies the* $f_B$ *part.*
2. $\overrightarrow{R}$'s *behaviour does not depend on anything* $\overrightarrow{S}$ *might modify: that is, for all* $a \in A$ *and for all* $c_1, c_2 \in C$, *if both* $f_A(c_1) = f_A(c_2)$ *and* $f_{rest}(c_1) = f_{rest}(c_2)$ *then*

$$f_A(\overrightarrow{R}(a,c_1)) = f_A(\overrightarrow{R}(a,c_2))$$

As suggested by the choice of terminology, we get

**Theorem 1.** *Let* $R : A \gtrless C$ *and* $S : B \gtrless C$ *be correct and hippocratic bx sharing a common target* $C$. *If there is a non-interfering* $A/B/rest$ *decomposition of* $C$, *then* $\overrightarrow{R}$ *and* $\overrightarrow{S}$ *are non-interfering.*

(For proof, see [7])

A straightforward special case illustrates the connections between the notions in this subsection and in the previous one:

**Corollary 1.** *Let $R : A \rightleftharpoons C$ and $S : B \rightleftharpoons C$ be correct, hippocratic and history-ignorant bx sharing a common target $C$, and suppose that $\sim_{RF}^{C} \equiv \sim_{SB}^{C}$ and $\sim_{RB}^{C} \equiv \sim_{SF}^{C}$. Then $\overrightarrow{R}$ and $\overrightarrow{S}$ are non-interfering.*

The proof is routine: we take the $A$ and $B$ parts to be given by the equivalences and the rest part to be trivial. History ignorance ensures that the conditions of Definition 5 hold – which is not a surprise, because history ignorance indicates that $C$ is morally a direct product of the part that is relevant to the other model with the part that is not.

## 4.3  Stamps, Slices and Optimality

The idea of a stamp on a model comes from the pluto framework where it is motivated by the wish to avoid unnecessary rebuilding of a target which, even though one or more of its sources have changed, does not need to be rebuilt because no source has changed in a way that is relevant to their target. It is originally given explicitly in terms of files and the file system, but for convenience we elide such details here, which causes a stamper to be formally almost identical to a part, although differently motivated.

**Definition 6.** *Given a model set $M$ and a set $S$ of stamp-values for $M$, a stamper is simply a function $s : M \longrightarrow S$.*

When a builder restores consistency to the model for which it is responsible, part of what it records is a current stamp value for each of the neighbouring models with which consistency was restored. The builder is responsible for the choice of stamper, but must use stamps fine enough to ensure that if a model changes without changing the stamp, consistency will not be lost. The framework uses this record to avoid invoking a builder that definitely has no work to do. The idea is that stamps should be quick to calculate and compare, and should reliably identify as many circumstances as possible when it is not necessary to do any consistency restoration work, such as applying a potentially expensive bx, because the non-change in the stamp guarantees that, even if the model has changed, it has not done so in a way that affects the model belonging to the builder that chose the stamp. The choice of stamper is an interesting problem: a very fine stamp (e.g. last-modified time) is cheap to use but it is clear that one can often do better in terms of saving unnecessary work. The use of the equivalences to provide and analyse stampers, and their use within networks of models, seems to hold promise in some circumstances, but is not explored here.

## 5   Conclusions

This paper has summarised the motivation behind, and some key points of, the author's recent work on networks of bidirectional transformations, and given pointers to further information. Much more remains to be done.

**Acknowledgments.** Conversations with too many people to list have informed this work, so let me just mention en masse the participants of Dagstuhl no. 18491 on *Multidirectional Transformations and Synchronisations* [1].

## References

1. Cleve, A., Kindler, E., Stevens, P., Zaytsev, V.: Multidirectional transformations and synchronisations (dagstuhl seminar 18491). Dagstuhl Rep. **8**(12), 1–48 (2018)
2. Dijkstra, E.W.: Selected Writings on Computing: A Personal Perspective, Chapter On the Role of Scientific Thought, pp. 60–66. Springer-Verlag, Heidelberg (1982)
3. Erdweg, S., Lichter, M., Weiel, M.: A sound and optimal incremental build system with dynamic dependencies. In: OOPSLA, pp. 89–106. ACM (2015)
4. Johnson, M., Stevens, P.: Confidentiality in the process of (model-driven) software development. In: Proceedings of the 7th International Workshop on Bidirectional Transformations, Bx 2018, co-located with 2nd International Conference on the Art, Science, and Engineering of Programming, ACM (2018)
5. Stevens, P.: Observations relating to the equivalences induced on model sets by bidirectional transformations. EC-EASST, 49 (2012)
6. Stevens, P.: Is bidirectionality important? In: Pierantonio, A., Trujillo, S. (eds.) ECMFA 2018. LNCS, vol. 10890, pp. 1–11. Springer, Cham (2018). https://doi. org/10.1007/978-3-319-92997-2_1
7. Stevens, P.: Maintaining consistency in networks of models: bidirectional transformations in the large. Softw. Syst. Model. **19**(1), 39–65 (2019)
8. Stevens, P.: Connecting software build with maintaining consistency between models: towards sound, optimal, and flexible building from megamodels. Softw. Syst. Model. **19**(4), 935–958 (2020)

# Bidirectional Collaborative Frameworks for Decentralized Data Management

Yasuhito Asano[1], Yang Cao[2(✉)], Soichiro Hidaka[3], Zhenjiang Hu[4,6], Yasunori Ishihara[5], Hiroyuki Kato[6], Keisuke Nakano[7], Makoto Onizuka[8], Yuya Sasaki[8], Toshiyuki Shimizu[9], Masato Takeichi[10], Chuan Xiao[8], and Masatoshi Yoshikawa[2]

[1] Toyo University, Tokyo, Japan
[2] Kyoto University, Kyoto, Japan
`yang@i.kyoto-u.ac.jp`
[3] Hosei University, Tokyo, Japan
[4] Peking University, Beijing, China
[5] Nanzan University, Nagoya, Japan
[6] National Institute of Informatics, Tokyo, Japan
[7] Tohoku University, Sendai, Japan
[8] Osaka University, Suita, Japan
[9] Kyushu University, Fukuoka, Japan
[10] University of Tokyo, Tokyo, Japan

**Abstract.** Along with the continuous evolution of data management systems for the new market requirements, we are moving from centralized systems towards decentralized systems, where data are maintained in different sites with autonomous storage and computation capabilities. There are two fundamental issues with such decentralized systems: *local privacy* and *global consistency*. By local privacy, the data owner wishes to control what information should be exposed and how it should be used or updated by other peers. By global consistency, the systems wish to have a globally consistent and integrated view of all data. In this paper, we report the progress of our BISCUITS (Bidirectional Information Systems for Collaborative, Updatable, Interoperable, and Trusted Sharing) project that attempts to systematically solve these two issues in distributed systems. We present a new bidirectional transformation-based approach to control and share distributed data, propose several distributed architectures for data integration via bidirectional updatable views, and demonstrate the applications of these architectures in ridesharing alliances and gig job sites.

## 1 Introduction

Along with the continuous evolution of data management systems for the new market requirements, we are moving from centralized systems, which had often led to vast and monolithic databases, towards decentralized systems, where data are maintained in different sites with autonomous storage and computation capabilities. A common practice is the collaboration or acquisition of companies:

© Springer Nature Switzerland AG 2022
G. Fletcher et al. (Eds.): SFDI 2021, CCIS 1457, pp. 13–51, 2022.
https://doi.org/10.1007/978-3-030-93849-9_2

there is large demand for different systems to be connected to provide valuable services to users, yet each company has its own goal and often builds its own applications and database systems independently without federating with others. As a result, we need to construct a decentralized system by integrating the independently-built databases through schema matching, data transformation, and update propagation from one database to another.

There are two fundamental issues with such decentralized systems, *local privacy* and *global consistency*. By local privacy, the owner of the data stored on a site may wish to *control* and *share* data by deciding what information should be exposed and how its information should be used and updated by other systems. By global consistency, the systems may wish to have a globally consistent view of all data, integrate data from different sites, perform analysis through queries, and update the integrated data.

**Local Privacy, Views, and Bidirectional Transformation.** Views play an important role in controlling access of data [19,21] and for sharing data of different schemas [17,22], since they were first introduced by Codd about four decades ago [9]. A view is a relation derived from base relations, which is helpful to describe dependencies between relations and achieves database security. Deeply associated with views is the classic *view update problem* [4,12]: given a view definition in the form of a query over base relations, the view update problem studies how to translate updates made to the view to updates to the original base relations.

On the other hand, researchers in the programming language community have generalized the view update problem to a general synchronization problem and designed various domain-specific languages to support so-called *bidirectional transformations* [6,20,30].

A bidirectional transformation (BX) consists of a pair of transformations: a forward and a backward transformation. The *forward* transformation $get(s)$ accepts a source $s$ (which is a collection of base relations in the setting of view updating), and produces a target view $v$, while the *putback* (backward) transformation $put(s, v)$ accepts the original source $s$ and an updated view $v$, and produces an updated source. These two transformations should be *well-behaved* in the sense that they satisfy the following round-tripping laws.

$$put(s, get(s)) = s \qquad \text{(GETPUT)}$$
$$get(put(s, v)) = v \qquad \text{(PUTGET)}$$

The GETPUT property (or Acceptability [4]) requires that no change on the view should result in no change on the source, while the PUTGET property (or Consistency [4]) demands that all changes to the view be completely translated to the source by stipulating that the updated view should be the one computed by applying the forward transformation to the updated source. The exact correspondence between the notion of well-behavedness in BX and the properties on view updates such as translation of those under a constant complement [4,12] has been extensively studied [47].

It has been demonstrated [7] that this language-based approach helps to solve the view update problem with a bidirectional query language, in which every query can be interpreted as both a view definition and an update strategy. However, the existing solution is unsatisfactory because the view update strategies are chosen at design time and hardwired into the language, and what users wish to express may well not be included in the set of strategies offered by the language.

**Global Consistency.** To share data globally, the current data sharing in the decentralized systems are based on either assuming a single shared global schema [25,63] or a peer-based approach without a shared schema [37,39,45]. The former is mainly designed for enterprise use cases, where data is exchanged between a small number of databases through a single global schema. The latter is intended for cases where there are many databases, so it is challenging to decide on a single global schema; instead, data is exchanged between peers (each manages its own database), and the update propagates cascadingly in the network. However, the problem of data sharing is not simple enough to fall into the above two categories. An ideal data sharing system should consider both aspects and combine both advantages of the two approaches.

Moreover, in the current decentralized data management systems, each peer can update its own database, and local consistency at each peer is ensured [34,37]. However, the update requests may contradict each other when integrated, and may violate the fine-grained access control and auditing requirements for collaborations and sharing [1], rendering existing techniques inapplicable for data management tasks that demand global consistency.

In this paper, we report our progress on solving these two issues in decentralized systems. In particular, we have conducted a 5-year national big project in Japan, called BISCUITS (Bidirectional Information Systems for Collaborative, Updatable, Interoperable, and Trusted Sharing)[1] since 2017, with researchers from National Institute of Informatics, Osaka University, Kyoto University, Nanzan University, Hosei University, Tohoku University and University of Tokyo. Our best wish is that this paper would provide a big picture of the research results, insights, and the new perspectives we have got during the project, paving the way for future further investigation.

The organization of the rest of the paper is as follows. We start with a new bidirectional transformation-based approach to controlling and sharing distributed data based on the view in Sect. 2. We then describe several architectures, including Dejima, BCDS agent, SKY, and OTMS Core for data integration between multiple database servers in Sect. 3. We discuss two applications of Dejima, i.e., ride-sharing alliances and gig job sites, in Sect. 4. Finally, we conclude the paper in Sect. 5.

---

[1] The project URL is http://www.prg.nii.ac.jp/projects/biscuits/project.html.

## 2    Bidirectional Transformation

The main difficulty in using views to control and share distributed data lies in the inherent ambiguity of view update strategies when given a view definition (or a forward transformation). We lack effective ways of controlling the view update strategy (or the putback transformation); it would be awkward and counterintuitive, if at all possible, to obtain our intended view update strategy by changing the view definition that is under our control, when the view definition becomes complicated.

We have taken it for granted that a view should be defined by a query and that a sound and intended update strategy should be automatically derived even if it is known that automatic derivation of an intended update strategy is generally impossible [38]. In this section, we present a new language-based approach to controlling and sharing distributed data based on the view: *A view should be defined through a view update strategy to the base relations rather than a query over them.* This new perspective is in sharp contrast to the traditional approaches, and it also gives a direction for solving the problem: view update strategies should be programmable.

In this section, we start with BIRDS, a new bidirectional transformation engine, where Datalog is a suitable and powerful language for specifying various view update strategies, present a novel algorithm for automatically deriving the unique view definition from a view update strategy, and explain how such updatable views can be efficiently implemented using the trigger mechanism in PostgreSQL. Then, we discuss the relationship among different bidirectional properties and its influence on the design of bidirectional languages. Finally, we show how to extend bidirectional transformation to multidirectional transformation.

### 2.1    BIRDS: A Bidirectional View Update Language

In this section, we briefly introduce our language-based approach [59–62] to view updates. Rather than designing new domain-specific languages for defining views as in the existing works [7,20], our approach allows programmers to use the Datalog language, a well-known query language, to completely specify how view updates are reflected in the source database. To reduce the users' burden in programming such a view update strategy, which is non-trivial and error-prone, we propose algorithms to validate, optimize, and debug the user-written Datalog programs. We further propose a novel mechanism for programmers to use predefined view update strategies in writing new ones, especially for recursively defined views. Our implemented system is integrated with a practical relational database management system by compiling Datalog programs of view update strategies into corresponding SQL statements.

**Programming View Update Strategies.** Consider a view $V$ defined by a query *get* over a source database $S$, as in other works [10,20], we formulate a view update strategy as a so-called putback transformation *put*, which takes as input the original source $S$ and an updated view $V'$ to produce a new source $S'$ as follows:

$$S \xrightarrow{\;get\;} V$$

The pair of *get* and *put* forms a bidirectional transformation [10]. Recall that $V$ is a single relation defined over a relational database $S$ that consists of source relations. Given an updated view relation and original source relations, a view update strategy (i.e., a putback transformation) specifies how to compute the new source relations, or equivalently, the corresponding updates for the original source relations. The following example illustrates our idea of using Datalog to specify such a view update strategy:

*Example 1.* Consider a source database $S$, which consists of two base binary relations, $r_1(X,Y)$ and $r_2(X,Y)$, and a view relation $v$ defined by a union over $r_1$ and $r_2$. The following is a Datalog program that describes an update strategy for the view.

(Rule 1)     $-r_1(X,Y) :- r_1(X,Y), \neg v(X,Y).$

(Rule 2)     $-r_2(X,Y) :- r_2(X,Y), \neg v(X,Y).$

(Rule 3)     $+r_1(X,Y) :- v(X,Y), \neg r_1(X,Y), \neg r_2(X,Y).$

Here, we use "+" and "−" preceding a relation symbol to denote insertion and deletion operations on the relation, respectively. The first two rules state that if a tuple $\langle X, Y \rangle$ is in $r_1$ or $r_2$ but not in $v$, it will be deleted from $r_1$ or $r_2$, respectively. The last rule states that if a tuple $\langle X, Y \rangle$ is in $v$ but in neither $r_1$ nor $r_2$, it will be inserted into $r_1$. Consider an original source instance $S = \{r_1(1,2), r_2(2,3), r_2(4,5)\}$ and an updated view $V = \{v(1,2), v(3,4), v(4,5)\}^2$, the result is $\Delta S = \{+r_1(3,4), -r_2(2,3)\}$ that means tuple $\langle 3, 4 \rangle$ is inserted into $r_1$ and tuple $\langle 2, 3 \rangle$ is deleted from $r_2$. By applying $\Delta S$ to $S$, we obtain a new source database $S' = \{r_1(1,2), r_1(3,4), r_2(4,5)\}$.

**Validation and Optimization.** To guarantee every view update is correctly propagated to the source, a view update strategy must be well-behaved in the sense that it satisfies the following *round-tripping* properties with the view definition *get*:

$$\forall\, S, \qquad put\,(S,\, get(S)) = S \qquad\qquad \text{(GetPut)}$$
$$\forall\, S,\, V', \qquad get\,(put\,(S,\, V')) = V' \qquad\qquad \text{(PutGet)}$$

---

² We write $r(a,b)$ to denote a tuple $\langle a, b \rangle$ in $r$.

The GETPUT property ensures that unchanged views correspond to unchanged sources, while the PUTGET property ensures that all view updates are completely reflected to the source such that the updated view can be computed again from the query *get* over the updated source.

More interestingly, for a view update strategy *put*, there is at most one view definition *get* that satisfies the round-tripping properties. Therefore, we define that a view update strategy is valid if the corresponding view definition exists.

We propose a validation algorithm that statically checks the validity of a view update strategy manually written in Datalog. Firstly, we check if the Datalog program is well-defined in the sense that there is no ambiguity in the source update results. In other words, there must be no insertion and deletion of the same tuple on the same relation. Secondly, we check if the corresponding view definition exists. Our validation algorithm reduces the checks to the satisfiability of Datalog queries.

The Datalog language is expected to not only be expressive enough to cover many view update strategies but also has good properties for deciding the validity of these Datalog programs. Here, we consider LVGN-Datalog [61] that is an extension of non-recursive Datalog with guarded negation, built-in predicates, constant and linear view predicate. LVGN-Datalog inherits the decidability of query satisfiability in guarded negation Datalog [5]. Our validation algorithm is both sound and complete for any view update strategies written in LVGN-Datalog.

Once the Datalog-written view update strategy is validated, it is guaranteed to produce no source update if the view is unchanged due to the GETPUT property. Only updates on the view lead to changes in the output of the Datalog program, and thus make the source updated. Therefore, we optimize the user-written Datalog program by transforming it into an incremental one that compute source updates from view updates more efficiently. Our incrementalization algorithm adopts the incremental view maintenance technique for Datalog introduced in [23].

**Debugging.** As presented previously, because of the required properties, writing view update strategies is error-prone. Considering Example 1, Rule 3 can be wrong written as follows: $-r_2(X,Y) :- r_2(X,Y), v(X,Y)$. This wrong rule results in a wrong source update: $\Delta S = \{-r_2(4,5)\}$ that deletes tuple $(4,5)$ from source relation $r2$, and thus makes the view and the source relations no longer consistent.

It is important to show to the programmer why the program is invalid and how to correct all the bugs. As presented previously, a valid view update strategy must satisfy the required properties for any instances of the source and updated view. Although the existence of a counterexample of the source and the updated view is sufficient to state the invalidity of the program, to illustrate the unexpected behaviour of the program, a counterexample must be realistic and friendly to the users. To generate such a counterexample, we create a symbolic instance of the view and source relations and transform the evaluation of the Datalog program with the expected property into a constraint program. By using a constraint solver, we obtain an interpretation of the symbolic input over which the expected property of the Datalog program is violated.

The unexpected behavior of the program can be observed through the execution of the program over a counterexample that the results are not as expected. However, there are many possible causes of such unexpected behavior, and thus it is very ambiguous to determine the detailed locations of bugs. Therefore, we allow the user to be involved in the debugging process and propose a navigation strategy to efficiently guide the users in locating the bugs in their programs with counterexamples. Specifically, we design a dialog-based user interface to allow user interaction and a debugging engine that exploits the provenance information of how a tuple is computed in the results of a Datalog program to resolve the ambiguity in locating bugs of the programs with the generated counterexamples.

**SQL Translation.** The Datalog program of a view update strategy makes the source updated whenever there are any view updates. To implement this reactive behaviour in practical database systems, we use the trigger mechanism [48]. Specifically, we compile the Datalog program into a SQL program that defines certain triggers and associated trigger procedures on the view. These trigger procedures are automatically invoked in response to view update requests, which can be any SQL statements of INSERT/DELETE/UPDATE. The triggers perform the following steps:

1. Handling update requests to the view to derive the corresponding delta relation of the view.
2. Checking the constraints if applying the delta relation from step (1) to the view.
3. Computing delta relations of the source and applying them.

**Recursive Views.** Our validation and debugging algorithms focus on views that are commonly defined by nonrecursive queries (e.g., queries in relational algebra) in traditional database management systems. However, when using relational databases to store complex data structures, e.g., trees and graphs, recursion is at the core of many queries [18]. A textbook example of a recursive query is graph transitive closure, which computes the transitive closure of a binary relation link(src,dst) that stores all the links in the graph. Another example is tree-like data structures such as XML or JSON documents that can be also represented as graphs, and thus can be stored in relations [57]. Recursive computation such as transitive closure is necessary for tree traversals.

Recursion makes Datalog expressive to represent view update strategies for views that are recursively defined. However, in comparison with nonrecursive views, writing update strategies for recursive views is a more expensive task for programmers.

Our approach [62] is to provide a variety of recursive Datalog rules that implement a certain number of recursive patterns over relations. These predefined recursive Datalog rules are parameterized so that they can be used by programmers to implement a new view update strategy. To guarantee the validity, we pre-validate the recursive Datalog rules and statically check the user-written program by our validation algorithm.

## 2.2 Possibilities of Replacement of BIRDS Queries

Asymmetric bidirectional transformations, which play an important role in our framework, need to satisfy a round-trip property called well-behavedness. In the current implementation of the Dejima architecture, all bi-directional transformations are written in the BIRDS (delta-Datalog) query language, and hence we can statically check for well-behavedness. However, the BIRDS query language has limited expressive power and may not be able to describe the bidirectional transformations desired by the user. In this section, we introduce the theory of bidirectional transformations, which is expected to be a basis for replacing the delta-Datalog language with the other programming languages.

There are two approaches for defining well-behaved bidirectional transformations. The first approach is to use general-purpose (i.e., Turing-complete) programming languages, in which any computable function can be defined. In this approach, the well-behavedness of bidirectional transformations needs to be manually checked or proved every time they are defined because the property is undecidable in general. In our project, we have investigated an essence of bidirectional transformations which may make it easier to confirm the well-behavedness and related properties. We inspected relations among 12 lens laws (Fig. 1) that have been proposed in literature for either alternative or auxiliary properties of bidirectional transformations [44]. The result is illustrated in Fig. 2 which can derive all of the Horn-style implications of the form $L_1 \wedge \cdots \wedge L_n \Rightarrow L$ with lens laws $L_1, \ldots, L_n$ and $L$.

$$\forall s, s' \in S, \quad put(s, get(s')) = s' \qquad \text{(STRONGGETPUT)}$$
$$\forall s \in S, \quad put(s, get(s)) = s \qquad \text{(GETPUT)}$$
$$\forall s \in S, \forall v \in V, \quad get(put(s, v)) = v \qquad \text{(PUTGET)}$$
$$\forall s \in S, \forall v, v' \in V, \quad put(put(s, v), v') = put(s, v') \qquad \text{(PUTPUT)}$$
$$\forall s \in S, \forall v \in V, \quad put(s, get(put(s, v))) = put(s, v) \qquad \text{(WEAKPUTGET)}$$
$$\forall s \in S, \forall v \in V, \quad put(put(s, v), get(put(s, v))) = put(s, v) \quad \text{(PUTGETPUT)}$$
$$\forall s \in S, \forall v \in V, \quad put(put(s, v), get(s)) = s \qquad \text{(UNDOABILITY)}$$
$$\forall s \in S, \forall v \in V, \quad put(put(s, v), v) = put(s, v) \qquad \text{(PUTTWICE)}$$
$$\forall s \in S, \exists v \in V, \quad put(s, v) = s \qquad \text{(SOURCESTABILITY)}$$
$$\forall s \in S, \exists s' \in S, \exists v \in V, \quad put(s', v) = s \qquad \text{(PUTSURJECTIVITY)}$$
$$\forall s, s' \in S, \forall v, v' \in V, \quad put(s, v) = put(s', v') \Rightarrow v = v' \qquad \text{(VIEWDETERMINATION)}$$
$$\forall s \in S, \forall v, v' \in V, \quad put(s, v) = put(s, v') \Rightarrow v = v' \qquad \text{(PUTINJECTIVITY)}$$

**Fig. 1.** 12 lens laws

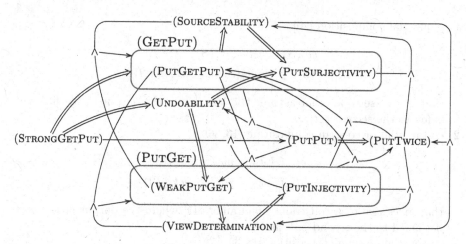

**Fig. 2.** Implication among lens laws

The second approach is to use the *BX languages* such as the BIRDS query language in which every program is either inevitably guaranteed or easily checked to define a well-behaved bidirectional transformation. A number of BX languages have been proposed since Boomerang [6], a combinator-based programming language, was proposed. However, it is not clear how 'expressive' the BX languages are, i.e., any computable bidirectional transformation can be defined as far as it is well-behaved. It would be nicer to have a computational model that exactly covers the set of computable well-behaved bidirectional transformations in order to prove whether each BX language is expressive enough. Towards this end, we have developed two computational models: an involutory Turing machine [42] and an idempotent Turing machine [43]. The involutory Turing machine is a computational model that can represent any computable involutory function, which is a function $f$ that is its own inverse, i.e., $f(f(x)) = x$ holds whenever $f(x)$ is defined. The idempotent Turing machine is a computational model that can represent any computable idempotent function, i.e., $f(f(x)) = f(x)$ holds whenever $f(x)$ is defined. These two models would give a hint for developing a computational model for well-behaved bidirectional transformations since the properties of involutoriness and idempotence are closely related to the well-behavedness [42].

## 2.3   Extension to Multidirectional Transformations

The idea of synchronization of more than two information sources by network of bidirectional transformations are well studied (for example, by Stevens [52]) as multidirectional transformations [8]. We can consider a pair of *get* and *put* of an asymmetric bidirectional transformation (lens) as a fundamental building block for such construction, and composition patterns can be classified into

1. sequential composition [20]

$$get(S) = get_2(get_1(S))$$
$$put(S, T) = put_1(S, put_2(get_1(S), T))$$

that gives rise to another asymmetric lens where the output of the first *get* is fed to the second *get*;

2. co-targetial or co-span composition [13,36]

$$put_{12}(S, T) = put_2(T, get_1(S))$$
$$put_{21}(S, T) = put_1(S, get_2(T))$$

that gives rise to a constraint maintainer [41] where the output of the first *get* is fed to the second *put*; and

3. co-sourcial or span composition [13,35]

$$put_{12}(S, C) = (get_2(C'), C')$$
$$\text{where } C' = put_1(C, S)$$
$$put_{21}(T, C) = (get_1(C'), C')$$
$$\text{where } C' = put_2(C, T)$$

that gives rise to a symmetric lens [31] where the output of the first *put* is fed to the second *get*. $C$ represents an internal state. In the above formulation, it is passed around as an additional argument but implementations could keep it entirely internal so that arguments are just $S$ and $T$ for each direction.

*Alignment of the Targets in the Co-targetial Composition.* Among the compositions described above, co-targetial compositions may be found non-trivial because the forward transformations are faced each other. In general, two functions $f : A \rightarrow B$ and $g : C \rightarrow D$ composes well if the range $B$ of $f$ is within the domain $C$ of $g$, i.e., $B \subseteq C$. When we consider both *get* and *put* directions, then, in order for the sequential compositions to be well-defined, the range of $get_1$ should be within the domain of $get_2$ and the first component of the domain of $put_1$ as well as the range of $put_2$ are within the second component of the domain of $put_1$. Therefore, the target of the first bidirectional transformation should match the source of the second bidirectional transformation. Similarly, the targets of the two bidirectional transformations should agree in the co-targetial composition, while the two sources should agree in the co-sourcial composition. The condition of co-targetial composition means the range of two *get* functions agree. If it is not the case, $put_2$ called in $put_{12}$ would fail. We could avoid such situation if we can align the two bidirectional transformations as follows, if the language that is used to describe the bidirectional transformation can compute the range in terms of predicates, like in relational lenses [7]. Suppose the target of the first and second bidirectional transformations are represented by the predicates $p_1$ and $p_2$, respectively, then the first bidirectional transformation can be composed with bidirectional projection transformation $\sigma_{p_2}$ and $\sigma_{p_1}$, respectively, so that the targets of the two new bidirectional transformations are both represented by the predicate $p_1 \wedge p_2$. Since the two target types become equal, the co-targetial composition is now well defined. This amendment effectively suppresses sending values that are out of the range of the get transformation of the counterpart bidirectional transformation.

In Sect. 3.2, the necessity to cope with more complex alignment and Smart Data Sharing solution using run-time amendment is discussed.

*Consistency Properties of Compositions.* These compositions maintain their own consistencies based on the notion of consistency of component lenses: a pair of source $s$ and target $t$ is consistent if $t = get(s)$. Well-behavedness of an asymmetric lens maintains this consistency as follows. If the source $s$ is modified to $s'$, then *get* transformation of $s'$ generates $t' = get(s')$ which is consistent with $s'$. If the target $t$ is modified to $t'$, then $put(s, t')$ generates $s'$ which is consistent with $t'$ by the PUTGET property.

Depending on the form of compositions, consistency is inherited from its components. A sequential composition maintains the same notion of consistency, since well-behavedness of its components implies well-behaved of the composite lens [20]. Similarly, a maintainer by co-targetial composition maintains consistency in the sense that the pair of $S$ and $put_{12}(S, T)$, as well as the pair of $T$ and $put_{21}(S, T)$, are consistent, if the component lenses are well-behaved. A symmetric lens by co-sourcial composition maintains consistency, given an internal state $C$ in the sense that the pair of $S$ and the first component of $put_{12}(S, C)$, as well as the pair of $T$ and the first component of $put_{21}(T, C)$, are consistent, if the component lenses are well-behaved.

Notably, a multispan composition that connects more than two bidirectional transformations in a star shape centering around an internal state gives rise to multiary lens [16]. The same argument applies for a multicospan composition [16].

*Versatility of Bipartite Graph of Asymmetric Bidirectional Transformations.* The Dejima architecture (described in detail in Sect. 3.2) can be considered as a bipartite graph [15] where Dejima groups and base tables constitute nodes, while asymmetric bidirectional transformations constitute edges, with base tables as their sources and Dejima tables as targets. Since *get* may discard information, co-targetial composition has been considered amenable to control exposure of information ([11], for example). From this viewpoint, the transformations between two peers connected via Dejima in an identical group can be considered as co-targetial composition and thus suitable for controlling information exposures from each peer. On the other hand, this composition is strictly less expressive than co-sourcial (span) composition [16], intuitively because the span composition can store every information necessary in the internal state $C$ to let the bidirectional transformations on both sides fully access the information. From this viewpoint, each peer can function as multiary lens [14] to support multidirectional transformation through multiple Dejimas and the internal state as its base table. Then we argue that, if we really need this expressiveness, we could replace each Dejima by another peer that is connected through each original Dejima. Figure 4 shows such case to replace Dejima $D_1$ in Dejima group $G_1$ of Fig. 3 by Dejima groups $G_{1S}$, $G_{2S}$ and $G_{3S}$, centered around a new dedicated peer $P_S$ responsible for synchronization among original Dejima group $G_1$.[3] From the viewpoint of the bipartite graph, the SKY architecture (explained in Sect. 3.4) can be represented by the Dejima architecture when we consider shared tables as base tables of dedicated peers as synchronizers, while base tables and their control tables as base tables and their Dejima tables.

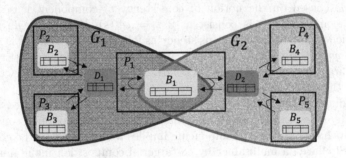

**Fig. 3.** An example of Dejima architecture: $P_n, G_n, D_n, B_n$ indicate peer, Dejima group, Dejima table, and base table, respectively.

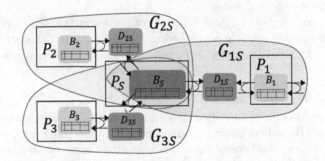

**Fig. 4.** More expressive bidirectional view updates achieved by replacing Dejima group $G_1$ depicted in Fig. 3 by Dejima groups $G_{1S}$, $G_{2S}$ and $G_{3S}$, centered around a new dedicated peer $P_S$ responsible for synchronization among original Dejima group $G_1$.

*Towards Concurrent Update Handling with the Help of Operational Transformations.* In the above discussions, we assumed that only one of two information sources is updated at the same time. For example, in case of a maintainer, $S$

---

[3] What is called Dejima table in Fig. 3 is expressed in Fig. 5 (Sect. 3.2) by two or more separate copies that belong to each peer.

side is authoritative when we conduct $put_{12}$. When we allow both ends to be updated in the bidirectional transformations, we need to consider more general synchronization scheme that handle concurrent update handling. A synchronizer sync : $S \times S \times T \times T \rightarrow S \times T$ instead of maintainer or symmetric lenses, can handle such situation. We propose in Sect. 3.5 the core part of such synchronizer based on existing work on operational transformation.

# 3    Bidirectional Collaborative Data Management

This section describes several architectures for data integration between multiple database servers (we call peers). In particular, we focus on an extended notion of data integration, *Bidirectional Collaborative Data Management*, in which multiple peers exchange database updates via bidirectional updatable views. We first give a summary of several architectures from major aspects and, then, describe the detail our contributions, the architectures of Dejima, BCDS agent, SKY, and OTMS Core in the following subsections. And then, we describe a key technique for recasting Schema Co-Existence for deployments of co-existent programs across evolving generations. It supports different versions of schemas of the common database each used by its corresponding program along with data synchronization, which is leveraged by bidirectional transformation.

## 3.1    Summary and Comparison on Architectures

We first summarize a history of Bidirectional Collaborative Data Management. Piazza [26,27] is one of the first projects on peer data management. Instead of using a single global schema, Piazza assumes semantic mappings between peers' individual schemas and introduces query reformulation between peers based on the mappings. PeerDB [45] is an integrated system of P2P system and data management system. It introduces several optimizations (configurable network, query result cache) for efficient keyword-based search over distributed peers. Both Piazza and PeerDB assumes read-only views for data exchange between peers. ORCHESTRA [34,37] is a successor project of Piazza. It permits update exchanges between peers (using updatable views) and supports autonomy on data management for each peer; each peer manages it own version of data. It also incorporates a novel model for tracking data provenance, such that each peer may filter updates based on trust conditions over this provenance.

Hereafter, we specifically focus on Bidirectional Collaborative Data Management, multiple peers exchange database updates via bidirectional updatable views. Table 1 gives an overview of the comparison between different architectures (ORCHESTRA and our work, Dejima, BCDS Agent, SKY, OTMS Core) from the following aspects.

**Data type.** Either relational database (SQL table), tree structure, or arbitrary data structure. For example, ORCHESTRA, Dejima, SKY are designed for relational databases, BCDS Agent is for arbitrary data structure, and OTMS Core is for tree structure.

**Peer composition by BX.** This aspect indicates how bidirectional transformation is composed for update exchange between peers. Dejima leverages co-targetial composition, BCDS Agent utilizes sequential/split/merge compositions, SKY and OTMS Core leverage co-sourcial composition. See Sect. 2.3 for more detail.

**Collaboration.** This aspects indicates the level (tightness or looseness) of collaboration between peers. In detail, the collaborate level is explained by the conformity level of updatable view specification between peers; how much $get(s)$ of one peer (export) is conformable to $put(s, v)$ (import) of another peer. For example, Dejima and BCDS Agent are designed in a tight level of collaboration in a sense that all the exported data from one peer are expected to be imported into adjacent peers. In contrast, in ORCHESTRA and SKY, a peer may reject importing data exported from its adjacent peers.

**Distributed consistency.** This aspects indicates a variation of distributed consistency (database synchronization) between peers. The strong consistency requires synchronous update propagation between peers (Dejima). The eventual consistency works with asynchronous update propagation (BCDS Agent, OTMS Core). In contrast, ORCHESTRA and SKY are designed based on multiversioning; they permit each peer to autonomously store its own version.

**Conflict resolution.** Some conflict resolution mechanism is necessary for ensuring the distributed consistency. The two-phase commit (2PC) and two-phase locking (2PL) is usually employed for ensuring strong consistency over distributed environments (Dejima). The eventual consistency requires convergence/confluence property of update propagation, such as provided by operational transformation (OT) [50] or conflict-free replicated data type (CRDT) [49]. For example, BCDS Agent employs first writer wins (FWW) for choosing an appropriate final database state when concurrent updates have occurred, whereas OTMS Core employs operational transformation for ensuring confluence of database state. In contrast, ORCHESTRA and SKY take a different approach; they optionally choose an appropriate final database state based on trust policies by leveraging data lineage/provenance.

We also note another important aspect, autonomy, which depends both on the collaboration and consistency aspects. The collaboration aspect affects compile-time autonomy based on the inconformity of view specification between peers. The consistency aspect affects execution-time autonomy based on database synchronization. For example, Dejima is less autonomous than the other architectures since it is designed with tight collaboration and strong consistency, whereas ORCHESTRA and SKY are highly autonomous since it is designed with loose collaboration and multi-versioning.

## 3.2 Dejima Architecture

The Dejima[4] architecture is our first BX-based architecture for achieving bidirectional collaborative data management. In the Dejima architecture, each peer

---

[4] Dejima was the name of a small, artificial island located in Nagasaki, Japan. All the trades between Japan and foreign countries were made through Dejima from the

collaborates with others through views called Dejima tables, which are defined by co-targetial composition of BX. Moreover, the Dejima architecture has transaction management mechanism over participating peers. As a result, the collaboration between peers realized by the Dejima architecture is quite close; The Dejima architecture connects peers at the database layer and the connected peers can behave as if they are using a single, same database. We call such style of collaboration *database collaboration*.

**Table 1.** Comparison on bidirectional collaborative data management architectures; any in the data type column indicates arbitrary data types. 2PC, 2PL, FWW, OT in the Conflict resolution column stand for two-phase commit, two-phase locking, first writer win, operational transformation, respectively.

| | Data type | Peer composition by BX | Collaboration | Distributed consistency | Conflict resolution |
|---|---|---|---|---|---|
| ORCHESTRA [34,37] | SQL Table | (Not applicable) | Loose | Multi-version | Lineage-base |
| Dejima [2,33] | SQL Table | Co-targetial | Tight | Strong | 2PC+2PL |
| BCDS Agent [54] | Any | Sequential, split, merge | Tight | Eventual | FWW |
| SKY (Sect. 3.4) | SQL Table | Co-sourcial | Loose | Multi-version | Lineage-base |
| OTMS Core [24] | Tree | Co-sourcial, co-targetial | Tight | Eventual | OT-base |

**Structure of the Dejima Architecture.** Suppose that two peers $P_i$ and $P_j$ agree to share subsets of their own data $B_i$ and $B_j$, respectively. In the Dejima architecture (Fig. 5), peer $P_i$ prepares a set $D_{ij}$ of Dejima tables, which is a set of views of $B_i$ to be shared with $P_j$ (and similarly, $P_j$ prepares $D_{ji}$). In the Dejima architecture, $P_i$ and $P_j$ collaborate through Dejima tables by maintaining $B_i$ and $B_j$ so that the equation $D_{ij} = D_{ji}$ holds.

To be more specific, $P_i$ prepares $D_{ij}$ by specifying the BX between $B_i$ and $D_{ij}$. Here, the BX represents what information of $P_i$ should be exposed to $P_j$ and how the information of $P_i$ should be updated by $P_j$. $D_{ij}$ and $D_{ji}$ must have the same schema. Then, $P_i$ and $P_j$ continue to update $B_i$ and $B_j$, respectively, so that the equation $D_{ij} = D_{ji}$ holds, according to their BX.

Now, we describe the Dejima architecture formally. Let $P_1, \ldots, P_n$ be participating peers, where each $P_i$ has a set $B_i$ of its base tables, that is, original tables owned by $P_i$. Let $D_{ij}$ be the set of Dejima tables from $P_i$ to $P_j$, where $D_{ij}$ and $D_{ji}$ have the same schema. Let $get_{ij}$ and $put_{ij}$ be the get and put functions

---

middle of the 17th to the middle of the 19th century. We use this name because the functionality is similar to Dejima.

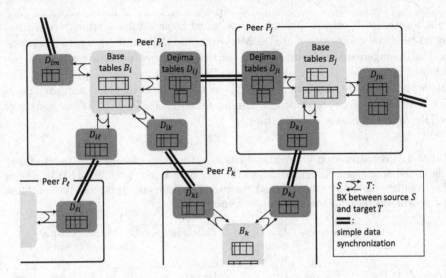

**Fig. 5.** The Dejima architecture [33]

between $B_i$ and $D_{ij}$. Then, from the round-tripping property, followings hold for any $P_i$ and $P_j$:

$$get_{ij}(B_i) = D_{ij}, \quad put_{ij}(B_i, D_{ij}) = B_i \tag{1}$$
$$get_{ji}(B_j) = D_{ji}, \quad put_{ji}(B_j, D_{ji}) = B_j \tag{2}$$

Note that the Dejima connection always keeps $D_{ij} = D_{ji}$ whenever changes in Dejima tables $D_{ij}$ or $D_{ji}$ occur. Suppose that $P_i$ updates the current $B_i$ to $B_i'$. Then the Dejima table $D_{ij}$ becomes $D_{ij}'$ and the equations

$$get_{ij}(B_i') = D_{ij}', \quad put_{ij}(B_i', D_{ij}') = B_i'$$

hold in $P_i$ from Eq. (1). Here, the Dejima connection keeps Dejima tables by yielding

$$D_{ij}' = D_{ji}'$$

and this causes $B_j$ to become $B_j'$ which satisfies

$$get_{ji}(B_j') = D_{ji}', \quad put_{ji}(B_j', D_{ji}') = B_j'$$

from Eq. (2). This process of equating the connected Dejima tables is called *Dejima synchronization*. Thus the above relation is invariant with respect to updates in peers' own data which are attained by the Dejima synchronization in the Dejima architecture. Note that each Dejima in Figs. 3 and 4 consists of Dejima tables connected by Dejima synchronization.

Update by a peer $P_i$ is propagated to peers indirectly connected to $P_i$. However, if one of such peers, say $P_j$, is processing another update in parallel, the data on $P_i$ and $P_j$ may be inconsistent. Or, as pointed out in [33], if the propagation forms a cycle, it may not terminate. To avoid such situations, the Dejima architecture has transaction management mechanism. As a result, the Dejima architecture achieves database collaboration, i.e., the database systems on the participating peers behave as if they are a single, same database as a whole. Hence, strong consistency of the data on different peers is guaranteed, and any database application running on a peer can be executed on other peers without any modification.

**Implementation of Dejima-Connected Systems.** We have implemented a platform called Dejima 1.0 based on the proposed architecture. In the platform, each peer uses PostgreSQL as its database management system. Strong consistency of the data is guaranteed by transaction management mechanism based on two-phase commit and two-phase locking. To improve the efficiency of parallel updates, we have introduced a novel data structure called *family record set* for distributed locking, which conservatively indicates a largest set of records to which a given update may propagated. The detail will be explained in [2] and the transaction management by leveraging distributed write ahead log is detailed in [3].

In what follows, we discuss several issues of the Dejima architecture.

*Initialization.* In general, $D_{ij}$ derived from $B_i$ and $D_{ji}$ derived from $B_j$ are different just after $P_i$ and $P_j$ agreed to collaborate with each other. We need some protocol for such initial synchronization of $D_{ij}$ and $D_{ji}$, depending on the application. One of the simplest protocols would be as follows: the initiator, say $P_i$, of the agreement changes the set $B_i$ of its base tables so that $D_{ij}$ becomes equal to $D_{ji}$.

*Materialization.* Interestingly, once the Dejima tables $D_{ij}$ and $D_{ji}$ are synchronized, they do not need to be materialized. To see this, suppose that $P_i$ has just made some update $\Delta B_i$ to $B_i$, where $\Delta B_i$ involves insertion, deletion, and modification. Using some operators informally, the update $\Delta D_{ij}$ to be made to $D_{ij}$ would be represented as:

$$\Delta D_{ij} = get_{ij}(B_i + \Delta B_i) - get_{ij}(B_i),$$

which can be computed from $B_i$ and $\Delta B_i$. Moreover, since $D_{ij} = D_{ji}$, the update $\Delta D_{ji}$ to be made to $D_{ji}$ must be equal to $\Delta D_{ij}$, which is sent from $P_i$ to $P_j$ through the synchronization mechanism. Hence, the update $\Delta B_j$ to be made to $B_j$ is:

$$\Delta B_j = put_{ji}(D_{ji} + \Delta D_{ji}, B_j) - B_j$$
$$= put_{ji}(get_{ji}(B_j) + \Delta D_{ij}, B_j) - B_j,$$

which can be computed from $B_j$ and $\Delta D_{ij}$. To ensure the efficiency of computing Dejima table deltas, e.g., $\Delta D_{ij}$, some well-known incrementalization techniques such as [23] may be applied.

**Smart Data Sharing.** We are too optimistic that we can always enjoy collaboration with peers by providing well-behaved BX according to individual policy for data sharing.

To achieve database collaboration through Dejima, the peer sometimes comes across problematic situations. It may fail to take complete autonomous actions. To be specific, the Dejima architecture assumes that

- The peer must accept any update coming from other peers through Dejima tables;
- For $P_i$ and $P_j$ to be connected, the range of $get_{ij}$ must be identical with that of $get_{ji}$; and
- What $P_i$ wants to export and what $P_j$ wants to import must be the same, and vise versa.

*Autonomy, Agreement and SDS.* Specifically, the second point above is worth noting that some *static agreement between $P_i$ and $P_j$* should be approved by the both peers before they are connected. However, in some cases, such a static agreement is appear to be not sufficient. SDS is a proposal for establishing a closer agreement between peers through flexible update exchange with BX-based views by managing view discrepancies between peers. Previous work on SDS [46] discusses a strategy that makes Dejima-connected peers manage collaboration by fixing incompatibility of their views on the way back and forth of update propagation; a peer can make negotiation on view details with the partner peer during the process of update exchange.

While this approach solves various problems for incompatible views provided by peers, SDS should cover more than that.

Remember that Dejima leverages the round-tripping property, which assures us of consistent view-updating between the view and the source in each peer. However, we are sometimes confronted with a basic issue related to BX-based view specification between peers. The cause leading to this problem comes from the fact that updating the view defined by BX allows us to use the source as well as the view itself for making them consistent, while it cannot use the source of the partner peer.

Not all the property and the structure of the view cannot be described in concise type definitions of programming languages or in SQL schemas for relational databases. Although as it is, the view specification of each peer can be associated with additional information as the *contract* based on the static agreement with peers. As a matter of course, the contract must be confirmed by calculation, or it must be defined by construction. Even with such contracts, a careful observation shows that updating either of the views sometimes fails to get the same instance of views when the contract depends on the source data as well as the view itself.

*The "Top-3" Example.* An example of such contracts appears in sharing the top-3 elements out of the source data. This could be easily agreed with peers without any ambiguity and it can be confirmed by calculation.

Consider an example in Fig. 6: initial source data of peers $P$ and $Q$ are sets $s_p = \{2, 8, 6, 4, 3\}$ and $s_q = \{4, 2, 6, 8, 1\}$, respectively. $get_p$ extracts top-3 largest numbers from $s_p$ to produce the view $v_p = \{8, 6, 4\}$. And similarly $get_q$ produces $v_q = \{8, 6, 4\}$ from $s_q$. Thus, the views facing each other are synchronized as $v_p = v_q = \{8, 6, 4\}$.

Here, we assume that $P$ or $Q$ update their source data with one of the operations by $P$ and $Q$ are "Replace $x$ with $x'''$", "Insert $x$ (if it does not exist in the source data)", or "Delete $x$", whichever cases that they change the source directly or they reflect the update propagated from the partner. And in the latter case, as a matter of course, though informally, the peer strives to apply as few operations as possible.

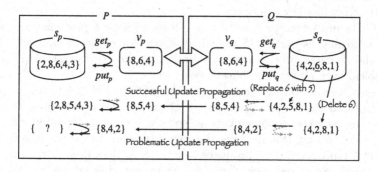

**Fig. 6.** Data sharing under the "top-3" contract

*Successful Update Propagation*: Peer $Q$ is about to update $s_q$ into $s_q' = \{4, 2, 5, 8, 1\}$ with the intention of replacing the element 6 by 5. This changes the view from $v_q = \{8, 6, 4\}$ to $v_q' = \{8, 5, 4\}$ by applying $get_q$ to $s_q'$. Next the updated view $v_q'$ is propagated to $v_p'$ for synchronization, which causes $v_p' = v_q' = \{8, 5, 4\}$. Note that $put_p$ is defined to accept the top-3 elements in the source set as the view, which satisfies the round-tripping property. So, the synchronized view $v_p'$ is propagated to the source for reflecting it into $s_p' = \{2, 8, 5, 4, 3\}$ as is expected. We can confirm the contract by calculating $get_p(s_p')$ to get the top-3 elements $\{8, 5, 4\}$. In fact, it is redundant because the round-tripping property ensures this.

*Problematic Update Propagation*: After the synchronization to share the top-3 elements of $P$ and $Q$ as $v_p = v_q = \{8, 6, 4\}$, what happens if $Q$ updates $s_q = \{4, 2, 6, 8, 1\}$ into $s_q' = \{4, 2, 8, 1\}$ by deleting the element 6? This update is followed by applying $get_q$ to $s_q'$ to get $v_q' = \{8, 4, 2\}$. Note again that the BX composed of $get_q$ and $put_q$ satisfies the round-tripping property, which we can confirm by calculating $v_q' = get_q(s_q')$ and $s_q' = put_q(s_q', v_q')$. In the course of update propagation, $v_q'$ is to be synchronized with $v_p$ to have the updated view $v_p' = \{8, 4, 2\}$. But this violates the round-tripping property between $get_p$ and $put_p$, whether we take this under reasonable semantics of this update: "6 is replaced by 2", or "6 is removed from the shared data". In either interpretation, the update, if allowed, causes to make $s'_p = \{2, 8, 4, 3\}$ from $s_p = \{2, 8, 6, 4, 3\}$, and $get_p(s'_p)$ produces $\{8, 4, 3\}$ as the top-3 view, which is not equal to $v'_p = \{8, 4, 2\}$.

*An SDS Solution.* Skeptical readers may offer alternative ways of updating the source; $put_p(s_p, v'_p) = v'_p \cup \{s \in s_p | \forall x \in v'_p, s < x\}$ applied to the above case does not cause the problem, nor even $put_p(s_p, v'_p) = v'_p$ does. Of course, when $P$ and $Q$ share these code as the contract, they are always happy to have successful results. It is, however, the Dejima architecture aims at peers' autonomous actions which are defined independently of other peers as described above. And that is where the SDS strategy comes in.

The SDS procedure proposed in [46] deals with such problematic cases by simply returning `Rejected`. Although it also enforces peer's updating policy during the repeated processes of view-updating, the above example illustrates the problem with already established contracts.

At least for the contract above and alike, SDS had better to take a proper action to fulfill the contract in stead of giving up synchronization. To do so, the first step is for $P$ to generously accept the temporary updated view $v'_p = \{8, 4, 2\}$ despite that this neglects the round-tripping property, or specifically the PUTGET property. This produces $s'_p = \{2, 8, 4, 3\}$ as described above. And then, $P$ produces a new view $v''_p = \{8, 4, 3\}$ by applying $get_p$ to $s'_p$. Note that the PUTGET property holds at this stage. This update can be considered as a new update of $P$'s source data to be propagated to $Q$, which will be accepted as the normal case satisfying the round-tripping at $Q$, or as the extended case under our generous SDS management.

While such patching can be considered as loosening BX's property from PUT-GET to PUTGETPUT, it remains for discussion to use such liberated BX in general for the Dejima architecture.

All in all, Smart Data Sharing extends the standard Dejima architecture to resolve view discrepancies in contracts between peers with well-behaved BXs.

## 3.3   BCDS Agent

The BCDS Agent [54] is a building unit for configuring collaborative data sharing systems with scalability and versatility. It rests on the novel feature of BX-based bidirectional programming [53] which encourages us to take the compositional approach in developing the distributed system with data consistency. We call a pair of forward and backward functions with bidirectional property *bidirectional function*.

**Bidirectional Function Composition.** We could construct a new bidirectional function from bidirectional functions using *combinators*. Constructing larger programs from smaller ones by combinators is a characteristic feature of Functional Programming. We assume here that bidirectional functions satisfy the properties GETPUT and PUTGETPUT throughout this section. While some composed function may satisfy stricter property, i.e., PUTGET, our combinators are more liberal in their bidirectionality. The combinators used in our BCDS Agent follow (Fig. 7).

*Bidirectional Sequential Composition.* The *sequential* composition $xf$<->$xg$ of bidirectional functions $xf$ and $xg$ is the most fundamental mechanism for constructing programs from primitive ones as the usual functional composition $f \cdot g$. Note that $xf$<->$xg$ applies forward $xf$ to the argument and then $xg$ to the result while $f \cdot g$ applies $f$ after $g$. The sequential composition keeps the bidirectional properties GETPUT and PUTGET and it has the transitivity property. These guarantee the consistency of the shared data among the related sites in the BCDS system.

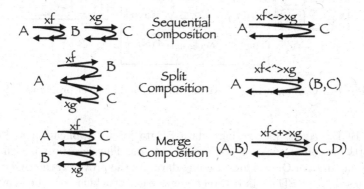

**Fig. 7.** Bidirectional function composition [54]

*Bidirectional Split Composition.* We use *split* composition[5] to make multiple target views from the source. While the split composition $xf$<^>$xg$ does not always inherit the PUTGET property of $xf$ and $xg$, it satisfies the PUTGETPUT property. We can find it in the BCDS System where the provider of the shared data may have multiple partner sites as receivers.

*Bidirectional Merge Composition.* We use *merge* composition $xf$<+>$xg$ to make multiple sources united by $xf$ and $xg$. It keeps the bidirectional properties GET-PUT and PUTGET of $xf$ and $xg$. In the BCDS system, the receiver gets data from the provider to make it available to the application. The shared data could be merged into a single data set by $xf$<+>$xg$ of $xf$ and $xg$ each transforms data from the corresponding provider.

---

[5] Split composition is a kind of composition patterns named *co-sourcial composition* explained in Sect. 2.3.

**Bidirectional Program for BCDS Agent.** In the general settings of data sharing systems, each participant site may work both as a provider supplying data to the others and as a receiver accepting data from others. The BCDS Agent possesses both the functions of *Provider* and *Receiver* with its proper *Application* program (Fig. 8).

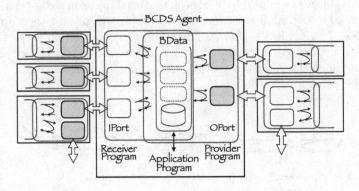

**Fig. 8.** BCDS agent: application, provider, and receiver programs [54]

The BCDS Agent has its base data BData for its application and several data ports for communication with other Agents. The data port is either IPort for incoming data or OPorts for outgoing data. As the provider, the BCDS Agent transforms data of BData into target views each of which is to be sent to the receiver through the corresponding OPort. And as the receiver, it accepts data sent from the provider with IPort to merge them into BData. The application program and the bidirectional programs for provider and for receiver run concurrently on separate threads of a machine with exclusive access to the base data BData. Bidirectional programs for the BCDS Agent represents the user's intention of sharing data with other Agents and they are developed by composing bidirectional functions with bidirectional combinators, i.e., sequential, split, and merge compositions..

*Bidirectional Provider Program.* We could construct multiple target views for OPorts from BData by split composition of bidirectional functions as far as only the one of the target views may be updated. Given bidirectional functions $xf$ and $xg$, split $xf<\hat{\ }>xg$ must satisfy the PUTGETPUT property for keeping the consistency of the source and the target views. The bidirectional program should evaluate $xf$ and $xg$ by applying them to the updated BData. So, either one of them may be discharged from this process because the target that caused this update is already consistent with the updated source due to the PUTGET property of $xf$ or $xg$. It is straightforward to extend this to the BCDS Agent with multiple target in general.

*Bidirectional Receiver Program.* We might consider that the receiver is the dual to the provider. But the construction of the bidirectional program for the receiver could not be done only by exchanging the source and the target of the bidirectional function. Rather than doing such, we write the bidirectional program using *Merge Composition* from multiple sources IPorts to a single target BData which originally consists of the local database and added the shared data. Since the target BData is the data set constructed as the discriminated union of the target data produced from the multiple source data, we can manage to put the update on BData back to IPorts with corresponding backward functions by necessity.

**BCDS System Configured with BCDS Agent.** We could build distributed Peer-to-Peer systems by connecting the BCDS Agent with the network. Collaborative tasks reside in the BCDS Agent with particular application and persistent local data.

*BCDS System Configuration.* The system has a directed graph structure which may be cyclic. Although we need to care about circular access to the shared data, we can detect and manage well because the BCDS model presupposes the ownership of the shared data and the provider sends the data to the receiver through connection. The provider always sends the data from its OPort to the IPort of the receiver and it also accepts the updated data backward from IPort to OPort. We do not need any provenance analysis more than identification of the ownership for detecting the self-referential data sharing [37]. We take this also in considering the propagation of updates in the BCDS system In each BCDS Agent, we should maintain BData, IPort data and OPort data to be consistent with each other whenever any change occurs. That is, every time when BData is changed, it is propagated to every partner site through corresponding IPort and OPort.

*Data Consistency Between Agents.* We should note that the distributed system configured with BCDS Agents keeps participants' BData consistent at least when no one is doing modification. This kind of the property called *eventual consistency* is ensured by the properties of the bidirectional program for the total system because it is composed of bidirectional programs in each participant Agent by bidirectional combinators.

*CDSS Simulation.* A BCDS system simulating a primitive part of the ORCHESTRA system [37] is built with three BCDS Agents as an simple example.

*Dejima Synchronizer.* In the Dejima architecture described above, synchronization between Dejima tables of peers is not a part of transformation, but is performed by the system itself, and it is beyond our control. With the BCDS Agent, the Dejima synchronizer peer can be built instead for carrying out various peers' policy on synchronization as the OTMS Core in Sect. 3.5 does. This realizes the separation of the transformation and the synchronization between collaborative peers.

*Real World Prototypes.* Several prototype systems are developed with BCDS Agents for examination including the Collaborative Taxi Dispatching Systems similar to the Ride-Sharing Alliances described in Sect. 4 and an experimental Turnaround Deliberation System.

## 3.4  SKY Architecture

The data management entities, that is peers, manage data autonomously, and may have demand for exchanging data with other peers based on local trust relationships. In such cases, it is desirable for each peer to be able to set policies for data import and export to prevent unintended data sharing. In addition, each peer does not necessarily maintain and publish the data in the same format.

The proposed system, which we call SKY, manages shared data with provenance information on the shared table (ST) in shared repository (SR), and each peer manages its data in its base tables (BTs) and controls the export and import of data through control table (CT).

An overview of the SKY architecture is shown in Fig. 9. Each peer participating in the sharing is assigned with a unique *pid*, and three peers, P1, P2, and P3, are sharing data. In addition, each peer can participate in multiple shares, and P2 shares data also with P4.

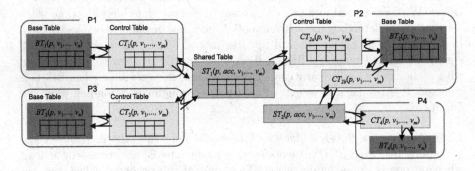

**Fig. 9.** Overview of the SKY architecture.

The SKY architecture utilizes BX to control import/export data from/to other peers. We introduce control tables (CTs) as views from base tables (BTs) and also as views from shared tables (STs) which contain all exported data from the peers participating in the system. We assume ST is stored in a shared repository which can be commonly accessed by peers, and it is considered as a dedicated peer for synchronization (See Sect. 2.3). In SKY, each peer can independently control importing data and exporting data. To this end, import policies (IPs) are defined as a view definition of CTs using STs whereas export policies (EPs) are defined as a view definition of CTs using BTs. Since we use BX in the view definitions, the view CT can be updatable so that update on BT can be propagated without side-effects to ST through CT, and vice versa.

In order to enable sharing policies using provenance, we introduce attribute $p$, which holds the provenance, to the BT, CT, and ST. Since there are cases in which we want to set a sharing policy that takes update patterns into account (e.g., not importing tuples that have been updated by other peers), we decided to define the provenance that takes into account the updates in addition to the general how-provenance. In the following, we describe the structure of the provenance and the functions involved in it. First, $p$ is defined recursively as follows.

$$
\begin{aligned}
p ::= \ & tid@pid && (* \text{ ground tuple } *) \\
| \ & p * p && (* \text{ join } *) \\
| \ & p + p && (* \text{ duplicates by union or projection } *) \\
| \ & p@pid && (* \text{ updated in pid } *)
\end{aligned}
$$

In addition, for convenience in defining policies, we define the functions $ulog$ and $org$ for $p$ with this structure.

$$
ulog(p) = \begin{cases}
[pid] & \text{if } p == tid@pid \\
[ulog(p1) \cup ulog(p2)] & \text{if } p == p1 * p2 \\
[ulog(p1) \cup ulog(p2)] & \text{if } p == p1 + p2 \\
(ulog(p1)) + [pid1] & \text{if } p == p1@pid1
\end{cases}
$$

$$
org(p) = \begin{cases}
\{pid\} & \text{if } p == tid@pid \\
(org(p1)) \cup (org(p2)) & \text{if } p == p1 * p2 \\
(org(p1)) \cup (org(p2)) & \text{if } p == p1 + p2 \\
org(p1) & \text{if } p == p1@pid1
\end{cases}
$$

We can get the update history sequence of $p$ by $ulog$, and the set of pids of the peers that originally owned the tuple by $org$. For example, $len(ulog(p)) == 1$ means that the tuple has not been updated by anyone. In this way, the provenance enables flexible and convenient policy description.

We also introduce the attribute $acc$ to the ST, which holds the set of pids of the peers that imported the tuple. For example, if a tuple is imported by two peers P2 and P3, the value of $acc$ is {P2, P3}. With $acc$, a peer can decide whether or not to import a tuple based on the import status of other peers. For example, $acc$ makes it possible to set the import policy of P1 to import tuples that are already imported by both P2 and P3.

## 3.5  OTMS Core

In this section, we propose OTMS Core (Operational Transformation-based Multidirectional Synchronization Core) to support concurrent update handling using operational transformations.

In Sect. 2.3, we mentioned the necessity for another synchronization scheme if we handle concurrent updates. Habu and Hidaka [24] worked on an instance of such situation by combining co-sourcial composition with operational transformation [50]. They extended the bidirectional duplication operator Dup in Hu et al.'s bidirectional transformation language X/Inv [32]. The original Dup duplicates the source in the forward transformation, while the updates on either view is reflected to the source in the backward transformation so that the following forward transformation propagates the updates to the other view. When both of the views are updated, the backward transformation is undefined. The extended Dup duplicates the source in the same way as the original in the forward direction, while in the backward direction, denoted by $\lhd'(s, s_1, s_2)$, it accepts concurrent updates on both views and reflects changes to the view by extracting operations on both views ($s$ to $s_1$ and $s$ to $s_2$, respectively), apply firstly one operation followed by the other operation transformed by the first one. The confluence property $C_1$ [50] of operational transformation guarantees that another order in which the other operation is applied firstly followed by the first operation transformed by the other one [24]. Using the extended operator $\mathsf{Dup}'$, we can build a synchronizer as follows. In co-sourcial composition, we start with a consistent pair of $t_1$ and $t_2$ such that $t_1 = get_1(s)$ and $t_2 = get_2(s)$ for some $s$. Then

$$\mathsf{sync}(t_1, t_1', t_2, t_2') = (t_1'', t_2'')$$
$$\text{where} \quad s_1' = put_1(s, t_1)$$
$$s_2' = put_2(s, t_2)$$
$$s' = \lhd'(s, s_1', s_2')$$
$$t_1'' = get_1(s')$$
$$t_2'' = get_2(s')$$

Consistency is restored because $t_1'' = get_1(s')$ and $t_2'' = get_2(s')$ for the new internal state $s'$. The synchronizer does not change the already consistent pair $(t_1', t_2')$ for the internal state $s$ such that $t_1' = get_1(s) = get_2(s) = t_2'$, if the component bidirectional transformations are well-behaved, because $put_1(s, t_1') = put_2(s, t_2') = s$ by GETPUT, and $s' = \lhd'(s, s, s) = s$. This property corresponds to the non-interference property in Song et al. [51]. Note that the original target $t_2$ is implicit in [51]. It would be ideal if the synchronizer outputs the argument as is if the argument consists of another consistent pair, i.e., if there exists another internal state $s'$ for a given pair $(t_1', t_2')$ such that $get_1(s') = get_2(s')$, then $\mathsf{sync}(t_1, t_1', t_2, t_2') = (t_1', t_2')$. This property corresponds to the stability property in Song et al. [51]. It is not always the case, regardless of the well-behavedness of the component bidirectional transformations, since $get$ may discard information so that more than one source value may correspond to one target. So $put_1(s, t_1')$ and $put_2(s, t_2')$ may be different. If both are the same, say, $s''$, even if different from $s'$, $\lhd'$ may lead to a value $s''$ as $\lhd'(s, s'', s'') = s''$, which would lead to the given consistent pair $(t_1', t_2')$ by PUTGET of component bidirectional transformations. Since it is not the case, $\lhd'$ is not guaranteed to produce a value $s'''$ such that $get_1(s''') = t_1'$ and $get_2(s''') = t_2'$. Consequently, stability is not satisfied in general by this synchronizer.

We can also build another synchronizer by co-targetial composition as follows. Starting with a consistent pair $(s_1, s_2)$ such that $get_1(s_1) = get_2(s_2) = t$,

$$\mathsf{sync}(s_1, s_1', s_2, s_2') = (s_1'', s_2'')$$
$$\text{where} \quad t_1' = get_1(s_1')$$
$$t_2' = get_2(s_2')$$
$$t' = \vartriangleleft'(t, t_1', t_2')$$
$$s_1'' = put_1(s_1', t')$$
$$s_2'' = put_2(s_2', t')$$

The synchronizer restores consistency if the component bidirectional transformations are well-behaved because $get_1(s_1'') = get_2(s_2'') = t'$ by PUTGET. This synchronizer can be used to initialize the Dejima for two peers having inconsistent base tables $B_1$ and $B_2$, by the empty initial Dejima table $D_0$ as the internal value $t$ for empty base tables $B_{10}$ and $B_{20}$, to set the new initial Dejima table to $t'$ and the new pair of base tables $(B_1', B_2') = \mathsf{sync}(B_{10}, B_1, B_{20}, B_2)$.

The synchronizer is non-interferent if the component bidirectional transformations are well-behaved because if $get_1(s_1) = get_2(s_2) = t$, then $\vartriangleleft'(t, t, t) = t$, and by the GETPUT property, $s_1 = put_1(s_1, t)$ and $s_2 = put_2(s_2, t)$. The synchronizer is stable if the component bidirectional transformations are well-behaved because if $get_1(s_1') = get_2(s_2') = t'$ for some $t'$, then $\vartriangleleft'(t, t', t') = t'$, and by the GETPUT property, $s_1' = put_1(s_1', t')$ and $s_2' = put_2(s_2', t')$.

When more than two information sources are involved, we need either the property $C_2$ for operational transformations, or we need a consistent way to coordinate transformations of operations to exhibit deterministic behavior of the synchronization. Such reasoning is left as our future work.

Once such multiary aspects are solved, we can extend the bipartite graph of bidirectional transformations described in Sect. 2.3 with operational transformations to support concurrent update handling, with evolutionary aspect in the configuration to replace the center of co-targetial composition with co-sourcial composition when we need more expressiveness in the synchronization, or degenerate a co-sourcial composition to a center of a co-targetial composition when such expressiveness is no longer needed.

## 3.6 Co-existence Schemas for Collaboration

While collaborative data management is most often discussed in the context of using databases distributed over the sites, it can also be considered in development of application programs which manipulate the common database and collaborate with each other across generations.

Schema evolution is an ever-occurring task inevitable in long-life data management systems. As systems evolve, they often need to update data schemas while maintaining backward and forward compatibility. Cambria [40] develops a library for schema evolution for distributed systems. Throughout the process, backward and forward compatibility should be maintained: backward compatibility in making new code compatible with existing data, and forward compatibility in making existing code compatible with new data.

From another aspect of schema evolution, we have to keep both application code for the old and the new schemas even after the new one becomes running. Although this is a kind of chronologically evolved applications using the same persistent data, both the schemas need to be co-existent for the backward and forward compatibility.

Also, schemas in peers connected through Dejima are evolved overtime in practice. In such a situation, both old and new schemas should be accessible so that existing bidirectional transformations between base tables and Dejima tables do not need to be changed. So, schema co-existence is needed to maintain Dejima network.

**Co-existent Schema Evolution.** As shown above, the schema co-existence is an important feature of a database. Unfortunately, the current relational database management systems do not support an efficient evolution of a schema and running multiple versions concurrently. As a notable work, Herrmann proposes MSVDB (Multi-Schema-Version Database Management System) that achieves a co-existence of relational schemas in the evolutional process of a database [28, 29]. Figure 10 shows a co-existence of two schemas. MSVDB composes a relational schema by views on top of shared physical data in one database. It provides a set of schema modification operations (SMOs) to show how to make an old schema evolved to a new one. It automatically gives a co-existence strategy for data sharing between these two schemas, i.e., a strategy for propagating updates between schemas through physical data.

**Bidirectional Schema Evolution for Co-existence.** Despite the user-friendly description of schema evolution and fully automatic mechanism, the MSVDB approach based on SMO has limited expressive power for users to describe co-existence strategies between two schemas. It prevents a user from solving many practical problems in a co-existence of schemas.

*Data Propagation Problem.* In the MSVDB approach, SMO describes a static relationship between the old and the new schemas, but it does not specify the strategy for data propagation (or data sharing) between the users of different schemas. The users may wish to share as many new information as possible even after evolution, or, to the other extreme, to do independently even if the two schemas co-exist. However, SMO cannot describe such an intended co-existing strategy; rather, it just provides a predefined strategy.

*Auxiliary Table Problem.* For the predefined strategy for each SMO given in the MSVDB approach, its implementation requires a tricky design of auxiliary tables that compensate supplemental information from physical data to an instance of a view. It lacks a systematical method to design these auxiliary tables, in which it is challenging to introduce new SMOs or change old SMOs even if we wish.

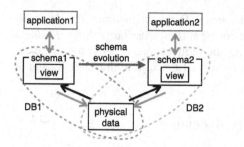

**Fig. 10.** Co-existence of two schemas

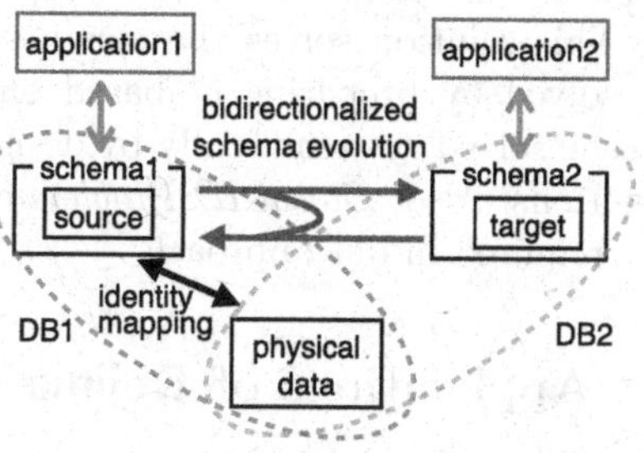

**Fig. 11.** Bidirectional schema evolution for co-existence

*Global-ID Problem.* The MSVDB approach requires a global id system among schemas before and after the evolution to record the correspondence among tuples for traceability. In practice, such a global id system is system-oriented, but not what the end-users wish to see. Moreover, the management of global id's uniqueness would cause performance bottleneck and unfavorable complexity of information systems.

We briefly present a BX-based solution to the above problems by proposing a new framework. We can found the details in [55,56].

To share updates on both old and new databases, a schema modification function is bidirectionalized such that the schema modification itself corresponds to a forward function and a strategy for update propagation from new databases to old ones corresponds to a backward function in bidirectional transformations. Figure 11 shows our approach. Here, we construct an identity mapping between a source (an old database) and a target (a new database), and carefully design the bidirectionalized schema modification function to realize the schema co-existence. In the schema co-existence, both sides of old and new behave like a full-fledged single-schema database in a sense that any update on each database can be accepted, and these update can be shared each other according to a schema evolution by a schema modification function.

To accept any update on a target, a totalized backward function is required in the bidirectionalized schema modification functions. Namely, for given a schema evolution function $f$ and an update strategy $b$, the extended schema modification function $(ft)$ and the totalized update propagation function $(bt)$ can be reformulated as follows:

$$ft(S \oplus A) = f(S) \oplus f'(A)$$
$$bt((S \oplus A), (V \oplus V')) = b(S, V) \oplus b'(A, V')$$

Where, $S$ is the data domain in an old database, $A$ is the domain of an auxiliary data to propagate out-of-range update on a new database, $V'$ is out-of-range update on the new database. Note that the similar approach to Smart Data Sharing was presented in Sect. 3.2.

This solution solves the existing problems. *Data Propagation Problem* can be solved by preparing *bt* based on users intention. *Auxiliary Table Problem* can be solved systematically by designing the bidirectional schema modification functions. Also, *Global-ID Problem* can be solved because a global id system is not required in our approach.

## 4 Applications of **Dejima** Architecture

In this section, we describe two applications for Dejima architecture: ride-sharing alliances and gig job sites.

### 4.1 Ride-Sharing Alliances

Here we describe the data sharing for ride-sharing alliances as an application of the Dejima architecture. Ride-sharing services allow non-professional drivers to provide taxi services using their vehicles. Each driver (or vehicle) usually belongs to a single ride-sharing provider such as Uber or Didi. As the ride-sharing market size increases, several ride-sharing service providers have formed partnerships with other providers to increase chances of matching providers to customers [58]. Actually, Uber and Didi have merged their China operations. Also, Ola Cabs and GrabTaxi are in talks to join a global taxi alliance.

In addition to ride-sharing, a variety of other service alliances are growing to increase their opportunities for matching customers. For instance, a marketplace platform such as e-Bay or Amazon marketplace can be regarded as an alliance of companies providing shopping services.

**Characteristics of Ride-Sharing Alliances.** We here explain why the ride-sharing alliances are a suitable application for database collaboration, even compared to other service alliances.

1. **Huge amount of data.** In 2016, there were more than 48,000 cabs and hire cars in Tokyo alone, transporting more than 262 million people per year; this number is likely to increase further as ride-sharing services like Uber become more popular. In fact, in New York City, the number of Uber drivers reportedly increased more than six-fold from 2015 to 2018. A database management system is needed for this huge amount of data.
2. **Impact of collaborative data sharing on services.** It is not uncommon for service providers to form alliances to expand their services to broad areas. For example, two Japanese taxi service providers Japan Taxi and MOV have formed the GO alliance recently. Data sharing is an essential part of establishing alliances. If the data sharing is badly designed, the individuality of each provider's data will be lost, and a large-scale redesign of their applications will be inevitable. Ideally, each service provider should be able to continue its services as before, and the application design of the alliance using the shared data should be as prepared as possible.

3. **Frequent transactions.** Transactions in a ride-sharing alliance occur not only when a vehicle is searched and assigned to each passenger, but also when the current location of a vehicle changes. Therefore, the number of transactions is proportional to the distance traveled, and according to the 2016 data for Tokyo mentioned above, the total distance traveled by all cabs and hire cars in a year exceeds 97 million kilometers, so the number of transactions is expected to be huge.

For these reasons, sharing data for ride-sharing alliances is extremely difficult at the application level, and requires architecture-level technologies such as the Dejima architecture.

**Ride-Sharing Alliances on Dejima.** We here explain the database collaboration between a provider and an alliance using the Dejima architecture. This database collaboration will be the basis of a demo application of ride-sharing alliances that we are implementing. We plan to include the detail of this implementation in a paper [2] that we are preparing for the submission. The content below including Figs. 12, 13, and 14 will be an excerpt from that paper.

Let us assume that each provider has a database that manages the data of vehicles belonging to the provider, and each alliance has the data of them; each provider can participate in multiple alliances and determine which vehicles are available on each alliance. Then, a pair of Dejimas is required for each pair of a provider and an alliance. Figure 12 illustrates an example of two alliances and three providers, and Dejimas between them.

Let us assume that each of Providers $A$ and $B$ in Fig. 12 has table bt whose attributes are V (the ID of a vehicle), L (its current location), D (its current destination), and R (the ID of a request that is assigned to it); and Alliance 1 has table mt that contains the data of shared vehicles published by $A$ and $B$, and attribute P to identify whether each vehicle belongs to the Provider $A$ or $B$. Then, when a passenger whose location is $\ell$ sends request $r$ to the Alliance 1, it selects available vehicles from the shared data, and assigns one of them to the passenger. Assuming that the selected one is vehicle $v$ of $A$. Then, the Alliance 1 updates its mt by setting R to $r$, and D to the location of the passenger, where V is $v$ and P is $A$. The update should be propagated to bt of $A$ so that the provider is able to know $v$ is assigned to the request $r$ and send the vehicle to $\ell$ to pick up the passenger.

The Dejima architecture achieves the database collaboration that enables such an update propagation between a provider and an alliance. For example, all $A$ has to do is to write a simple BIRDS (see Sect. 2.1) code described in Fig. 13, while the Alliance 1 should write a BIRDS code shown in Fig. 13 for $A$. Similar BIRD codes should be written for each pair of a provider and an alliance. Then, the Dejima architecture guarantees the strong global consistency among all the tables of the providers and the alliances automatically.

**Hierarchical Alliances and Problem.** We here discuss the data sharing of hierarchical alliances. In the beginning of this subsection, we introduced a

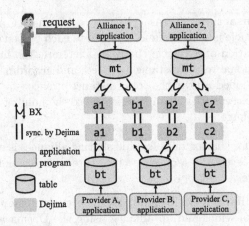

**Fig. 12.** Dejimas for ride-sharing alliances.

```
source bt('V':int,'L':int,'D':int,'R':int).
view a1('V':int,'L':int,'D':int,'R':int).
%view definition
a1(V,L,D,R) :- bt(V,L,D,R).
%update strategies
-bt(V,L,D,R) :- bt(V,L,D,R), NOT a1(V,L,D,R).
+bt(V,L,D,R) :- NOT bt(V,L,D,R), a1(V,L,D,R).
```

**Fig. 13.** BIRDS code between Provider A and Dejima a1.

fact that JapanTaxi and MOV have formed alliance GO. Actually, both the JapanTaxi and MOV are alliances in which several providers participate. Thus, GO is considered as an alliance of alliances. In this way, there is a possibility in various service alliances that multiple alliances providing localized services reorganize into a larger alliance in order to expand their services to a broader area. Then, services alliances can be a hierarchical structure.

Figure 15 illustrates an example of the hierarchical structure that consists of five providers, four alliances, and two "large" alliances. AL1, AL2, AL3, and AL4 represent Alliance 1, 2, 3 and 4, respectively. Each provider participates in one or two alliances in this example. For example, Provider $C$ participates in the Alliances 1 and 2. A large alliance has the data of vehicles of several alliances as

```
source mt('V':int,'L':int,'D':int,'R':int,'P':string).
view a1('V':int,'L':int,'D':int,'R':int).
%update strategies
-mt(V,L,D,R,P) :- mt(V,L,D,R,P), NOT a1(V,L,D,R), P='A'.
+mt(V,L,D,R,P) :- NOT mt(V,L,D,R,P), a1(V,L,D,R), P='A'.
```

**Fig. 14.** BIRDS code between Alliance 1 and Dejima a1.

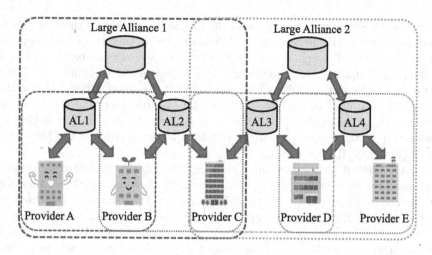

**Fig. 15.** Hierarchical model.

if they are providers participating in it, and receives requests from passengers. In this example, Large Alliance 1 (or 2) has the vehicle data of the Alliances 1 and 2 (or Alliances 3 and 4, respectively). Due to space limitations, we omit passengers and their requests in this figure.

As mentioned above, such a hierarchy would appear on various kinds of data sharing applications. However, we have found a problem of the hierarchical alliances for the collaborative data sharing architectures.

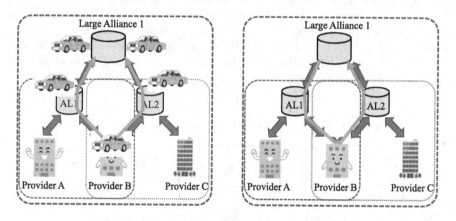

**Fig. 16.** (a) an example of a duplicate, and (b) an example of a cycle.

Figure 16 portrays two problems of the hierarchical alliances: a duplicate problem and a cycle problem. Let us assume that all the vehicles on the Alliances 1 and 2 are available on the Large Alliance 1.

Figure 16(a) depicts an example of the duplicate problem. If the Provider $B$ makes a vehicle available on both the Alliances 1 and 2, then this vehicle appears on the databases of them. As a result, this vehicle appears twice on the database of the Large Alliance 1. This problem can be solved on the application level. For example, assuming that a vehicle on the table of the Large Alliance 1 has the ID of a vehicle used in each provider, the ID of a provider, and the ID of an alliance. Then, the application of the Large Alliance 1 can identify a vehicle $v_1$ obtained from the Alliance 1 with $v_2$ obtained from the Alliance 2 if they have the same vehicle ID and provider ID.

Figure 16(b) depicts an example of the cycle problem. If the Large Alliance 1 updates the data of the vehicle of the Provider $B$ explained above, then the update should be propagated in the following order: the Alliance 1, the Provider $B$, the Alliance 2, and then the update reaches the Large Alliance 1 again. It is known that such a cycle is problematic on update propagation. For example, the Dejima architecture cancels a transaction if a cycle is found through its update propagation. Solving this problem on the architecture level would be one of the important future tasks.

## 4.2  Gig Job Sites

As another example of database collaboration using Dejima architecture, we consider gig job sites. Each gig job site collects and posts its job listing on a private page which is accessible only from their own registered workers. These sites earn commission when the matching between a job and a worker is established. It might be difficult, however, for a single site to find workers of certain types of jobs. Hence, the sites agree to form an alliance by sharing part of job data so that such data becomes visible to a larger set of workers. We can use Dejima architecture for this purpose. Figure 17 shows a configuration of gig job sites using Dejima architecture. Each site $P_i$ has its own job listing record $R_i$ and share a subset of $R_i$ with other sites. In this case, a dejima $D$ is a set of such shared records. Also, the base table $B_i$ of $P_i$ is the union of $R_i$ and $D$.

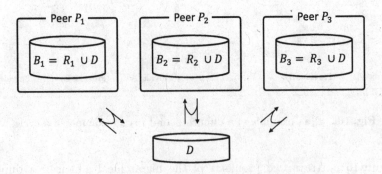

**Fig. 17.** A configuration of gig job sites.

$R_1$

| pid | jid | skill | due | open |
|-----|-----|-------|-----|------|
| $p_1$ | $j_1$ | Java | 60 | N |
| $p_1$ | $j_2$ | C++ | 3 | Y |

$R_2$

| pid | jid | skill | due | open |
|-----|-----|-------|-----|------|
| $p_2$ | $j_3$ | SQL | 14 | Y |
| $p_2$ | $j_4$ | SQL | 30 | Y |

$R_3$

| pid | jid | skill | due | open |
|-----|-----|-------|-----|------|
| $p_3$ | $j_5$ | Java | 7 | Y |
| $p_3$ | $j_6$ | SQL | 21 | Y |

**Fig. 18.** Gig job site databases.

We show a simple example of tables $R_1$, $R_2$, and $R_3$ in Fig. 18. For simplicity, we assume that all the tables have the same schema with five attributes p(eer_)id, j(ob_)id, skill, due(_period) and open, though each site may have its own schema in a real-world application. The value of due represents days, and the value of open is "Y" if the job is currently open, and "N" otherwise.

In this example, $P_1$ and $P_3$ respectively share the job data $j_2$ and $j_6$ among the three sites. This is because the due period of $j_2$ is only three days. Meanwhile the job site $P_3$ may find it is difficult to find a worker for the job $j_6$, because $P_3$ has only a small number of SQL programmers registered on its site. That is, the three sites share the data $D$ in Fig. 19.

Each site now has job data $D$ in addition to its own job data $R_i$ ($i = 1, 2, 3$). Here we have the following observations:

1. Applications in site $P_i$ should have an access to job data $R_i \cup D$ as a single monolithic table rather than as two separate tables $R_i$ and $D$ each stored in different databases.
2. $D$ can be regarded as a view of $B_i = R_i \cup D$.
3. An update on a data in $D(\subseteq B_i)$ must be propagated to $B_j, j \neq i$. For example, if $P_3$ founds a worker for the job $j_6$, $P_3$ changes the open value of $j_6$ in $B_3$ to "N". This update is automatically reflected to the shared data $D$ because $D$ is a view of $B_3$. The update on $D$, in turn, is propagated to $B_j, j \neq 3$. We can regard this propagation from $D$ to $B_1$ and $B_2$ as view update.

$D$

| pid | jid | skill | due | open |
|-----|-----|-------|-----|------|
| $p_1$ | $j_2$ | C++ | 3 | Y |
| $p_3$ | $j_6$ | SQL | 21 | Y |

**Fig. 19.** A Dejima table for gig job sites.

## 5   Conclusion

This paper reported our progress on systematically solving the local privacy and global consistency issues in distributed systems. We proposed a new language-based approach to control and share distributed data based on the view. Our insight is that a view should be defined through a view update strategy to the base relations rather than a query over them. This new perspective is in sharp contrast to the traditional approaches, and it also provides a direction for solving the problem: view update strategies should be programmable. To substantiate our perspective, we designed four architectures, i.e., Dejima, BCDS agent, SKY, and OTMS Core, for data integration between multiple database servers to support diverse data types, peer composition by BX, collaboration styles, distributed consistency, and conflict resolution strategies. Further, we demonstrated the applications with the featured Dejima architecture.

We believe that it is worth reporting as early as possible the new perspective on views and the view-based programmable data management architecture arising from it, so that researchers in databases and programming languages can start working together to explore this promising direction.

**Acknowledgments.** This work was partly supported by JSPS KAKENHI Grant Numbers 17H06099, 18H04093, 19H04088. We also thank Dr. Zinovy Diskin (McMaster University) for discussing bipartite graph representation and expressiveness of the Dejima architecture.

# References

1. Abadi, D., et al.: The Seattle report on database research. ACM SIGMOD Rec. **48**(4), 44–53 (2020)
2. Asano, Y., et al.: Dejima: decentralized transactional data integration with bidirectional update propagation. To be Submitted to an International Conference
3. Asano, Y., et al.: Flexible framework for data integration and update propagation: system aspect. In: 2019 IEEE International Conference on Big Data and Smart Computing (BigComp), pp. 1–5 (2019)
4. Bancilhon, F., Spyratos, N.: Update semantics of relational views. ACM Trans. Database Syst. **6**(4), 557–575 (1981)
5. Bárány, V., ten Cate, B., Otto, M.: Queries with guarded negation. PVLDB **5**(11), 1328–1339 (2012)
6. Bohannon, A., Foster, J.N., Pierce, B.C., Pilkiewicz, A., Schmitt, A.: Boomerang: resourceful lenses for string data. In: POPL, pp. 407–419 (2008)
7. Bohannon, A., Pierce, B.C., Vaughan, J.A.: Relational lenses: a language for updatable views. In: PODS, pp. 338–347 (2006)
8. Cleve, A., Kindler, E., Stevens, P., Zaytsev, V.: Multidirectional transformations and synchronisations (Dagstuhl seminar 18491). Dagstuhl Rep. **8**(12), 1–48 (2019)
9. Codd, E.F.: Recent investigations in a relational database system. Inf. Process. **74**, 1017–1021 (1974)
10. Czarnecki, K., Foster, J.N., Hu, Z., Lämmel, R., Schürr, A., Terwilliger, J.F.: Bidirectional transformations: a cross-discipline perspective. In: Paige, R.F. (ed.) ICMT 2009. LNCS, vol. 5563, pp. 260–283. Springer, Heidelberg (2009). https://doi.org/10.1007/978-3-642-02408-5_19
11. Dampney, C.N.G., Johnson, M.: Half-duplex interoperations for cooperating information systems. In: Advances in Concurrent Engineering, pp. 565–571 (2001)
12. Dayal, U., Bernstein, P.: On the correct translation of update operations on relational views. ACM Trans. Database Syst. **7**, 381–416 (1982)

13. Diskin, Z.: Algebraic models for bidirectional model synchronization. In: Czarnecki, K., Ober, I., Bruel, J.-M., Uhl, A., Völter, M. (eds.) MODELS 2008. LNCS, vol. 5301, pp. 21–36. Springer, Heidelberg (2008). https://doi.org/10.1007/978-3-540-87875-9_2
14. Diskin, Z.: Update propagation over a network: multi-ary delta lenses, tiles, and categories. Keynote Talk of the 3rd Workshop on Software Foundations for Data Interoperability (SFDI 2019+) (2019)
15. Diskin, Z., Hidaka, S.: Personal communications, October 2019
16. Diskin, Z., König, H., Lawford, M.: Multiple model synchronization with multiary delta lenses. In: Russo, A., Schürr, A. (eds.) FASE 2018. LNCS, vol. 10802, pp. 21–37. Springer, Cham (2018). https://doi.org/10.1007/978-3-319-89363-1_2
17. Doan, A., Halevy, A.Y., Ives, Z.G.: Principles of Data Integration. Morgan Kaufmann, Waltham (2012)
18. Dong, G., Su, J.: Incremental maintenance of recursive views using relational calculus/SQL. SIGMOD Rec. **29**(1), 44–51 (2000)
19. Fernandez, E.B., Summers, R.C., Wood, C.: Database Security and Integrity. Addison-Wesley, Reading (1981)
20. Foster, J.N., Greenwald, M.B., Moore, J.T., Pierce, B.C., Schmitt, A.: Combinators for bidirectional tree transformations: a linguistic approach to the view-update problem. ACM Trans. Program. Lang. Syst. **29**(3), 17 (2007)
21. Foster, J.N., Pierce, B.C., Zdancewic, S.: Updatable security views. In: CSF, pp. 60–74 (2009)
22. Golshan, B., Halevy, A., Mihaila, G., Tan, W.-C.: Data integration: after the teenage years. In: PODS, pp. 101–106 (2017)
23. Gupta, A., Mumick, I.S., Subrahmanian, V.S.: Maintaining views incrementally. In: SIGMOD, pp. 157–166 (1993)
24. Habu, M., Hidaka, S.: Conflict resolution for data updates by multiple bidirectional transformations. In: Proceedings of the Fifth Workshop on Software Foundations for Data Interoperability (SFDI 2021), August 2021. (to appear)
25. Halevy, A.Y.: Answering queries using views: a survey. VLDB J. **10**(4), 270–294 (2001). https://doi.org/10.1007/s007780100054
26. Halevy, A.Y., Ives, Z.G., Madhavan, J., Mork, P., Suciu, D., Tatarinov, I.: The Piazza peer data management system. IEEE Trans. Knowl. Data Eng. **16**(7), 787–798 (2004)
27. Halevy, A.Y., Ives, Z.G., Mork, P., Tatarinov, I.: Piazza: data management infrastructure for Semantic Web applications. In: WWW, pp. 556–567 (2003)
28. Herrmann, K., Voigt, H., Behrend, A., Rausch, J., Lehner, W.: Living in parallel realities: co-existing schema versions with a bidirectional database evolution language. In: Proceedings of the 2017 ACM International Conference on Management of Data, pp. 1101–1116 (2017)
29. Herrmann, K., Voigt, H., Pedersen, T., Lehner, W.: Multi-schema-version data management: data independence in the twenty-first century. VLDB J. **27**(4), 547–571 (2018). https://doi.org/10.1007/s00778-018-0508-7
30. Hidaka, S., Hu, Z., Inaba, K., Kato, H., Matsuda, K., Nakano, K.: Bidirectionalizing graph transformations. In: ICFP, pp. 205–216 (2010)
31. Hofmann, M., Pierce, B.C., Wagner, D.: Symmetric lenses. In: POPL, pp. 371–384 (2011)
32. Hu, Z., Mu, S.-C., Takeichi, M.: A programmable editor for developing structured documents based on bidirectional transformations. Higher-Order Symb. Comput. **21**(1–2), 89–118 (2008). https://doi.org/10.1007/s10990-008-9025-5

33. Ishihara, Y., Kato, H., Nakano, K., Onizuka, M., Sasaki, Y.: Toward BX-based architecture for controlling and sharing distributed data. In: 2019 IEEE International Conference on Big Data and Smart Computing (BigComp), pp. 1–5 (2019)

34. Ives, Z., Khandelwal, N., Kapur, A., Cakir, M.: ORCHESTRA: rapid, collaborative sharing of dynamic data. In: CIDR, pp. 107–118 (2005)

35. Johnson, M., Rosebrugh, R.D.: Spans of lenses. In: Candan, K.S., Amer-Yahia, S., Schweikardt, N., Christophides, V., Leroy, V. (eds.) CEUR@EDBT/ICDT, pp. 112–118 (2014)

36. Johnson, M., Rosebrugh, R.D.: Cospans and symmetric lenses. In: Marr, S., Sartor, J.B. (eds.) Conference Companion of the 2nd International Conference on Art, Science, and Engineering of Programming, pp. 21–29 (2018)

37. Karvounarakis, G., Green, T.J., Ives, Z.G., Tannen, V.: Collaborative data sharing via update exchange and provenance. ACM Trans. Database Syst. **38**(3), 19:1–19:42 (2013)

38. Keller, A.: Choosing a view update translator by dialog at view definition time. In: VLDB, pp. 467–474 (1986)

39. Kementsietsidis, A., Arenas, M., Miller, R.J.: Mapping data in peer-to-peer systems: semantics and algorithmic issues. In: SIGMOD, pp. 325–336 (2003)

40. Litt, G., van Hardenberg, P., Henry, O.: Cambria: schema evolution in distributed systems with edit lenses. In: Proceedings of the 8th Workshop on Principles and Practice of Consistency for Distributed Data (PaPoC 2021), pp. 1–9. ACM Digital Library (2021). Article no. 8

41. Meertens, L.: Designing constraint maintainers for user interaction (1998). http://www.kestrel.edu/home/people/meertens

42. Nakano, K.: Involutory turing machines. In: Lanese, I., Rawski, M. (eds.) RC 2020. LNCS, vol. 12227, pp. 54–70. Springer, Cham (2020). https://doi.org/10.1007/978-3-030-52482-1_3

43. Nakano, K.: Idempotent turing machines. In: Bonchi, F., Puglisi, S.J. (eds.) 46th International Symposium on Mathematical Foundations of Computer Science, MFCS 2021, Volume 202 of LIPIcs, Tallinn, Estonia, 23–27 August 2021, pp. 79:1–79:18. Schloss Dagstuhl - Leibniz-Zentrum für Informatik (2021)

44. Nakano, K.: A tangled web of 12 Lens laws. In: Yamashita, S., Yokoyama, T. (eds.) RC 2021. LNCS, vol. 12805, pp. 185–203. Springer, Cham (2021). https://doi.org/10.1007/978-3-030-79837-6_11

45. Ng, W.S., Ooi, B.C., Tan, K.-L., Zhou, A.: PeerDB: a P2P-based system for distributed data sharing. In: ICDE, pp. 633–644 (2003)

46. Onizuka, M., Ishihara, Y., Takeichi, M.: Towards smart data sharing by updatable views. In: Qin, L., et al. (eds.) SFDI/LSGDA -2020. CCIS, vol. 1281, pp. 165–171. Springer, Cham (2020). https://doi.org/10.1007/978-3-030-61133-0_13

47. Pierce, B.C., Schmitt, A.: Lenses and view update translation. Working draft, University of Pennsylvania (2003)

48. Ramakrishnan, R., Gehrke, J.: Database Management Systems, 2nd edn. McGraw-Hill Inc., New York (1999)

49. Shapiro, M., Preguiça, N., Baquero, C., Zawirski, M.: Conflict-free replicated data types. In: Défago, X., Petit, F., Villain, V. (eds.) SSS 2011. LNCS, vol. 6976, pp. 386–400. Springer, Heidelberg (2011). https://doi.org/10.1007/978-3-642-24550-3_29

50. Sinchuk, S., Chuprikov, P., Solomatov, K.: Verified operational transformation for trees. In: Blanchette, J.C., Merz, S. (eds.) ITP 2016. LNCS, vol. 9807, pp. 358–373. Springer, Cham (2016). https://doi.org/10.1007/978-3-319-43144-4_22

51. Song, H., et al.: Supporting runtime software architecture: a bidirectional-transformation-based approach. J. Syst. Softw. **84**(5), 711–723 (2011)
52. Stevens, P.: Bidirectional transformations in the large. In: MODELS, pp. 1–11 (2017)
53. Takeichi, M.: Configuring bidirectional programs with functions. In: Draft Proceedings of the 21st International Symposium on Implementation and Application of Functional Languages, pp. 224–239 (2009)
54. Takeichi, M.: BCDS agent: an architecture for bidirectional collaborative data sharing. In: Computer Software, vol. 38, no. 3, pp. 41–57. Japan Society for Software Science and Technology (2021). https://www.jstage.jst.go.jp/article/jssst/38/3/38_3_41/_pdf/-char/ja
55. Tanaka, J., Tran, V.-D., Hu, Z.: Toward programmable strategy for co-existence of relational schemes. In: Qin, L., et al. (eds.) SFDI/LSGDA-2020. CCIS, vol. 1281, pp. 138–151. Springer, Cham (2020). https://doi.org/10.1007/978-3-030-61133-0_11
56. Tanaka, J., Tran, V.-D., Kato, H., Hu, Z.: Toward co-existing database schemas based on bidirectional transformation. In: Proceedings of the 3rd Workshop on Software Foundations for Data Interoperability (SFDI 2019+) (2019)
57. Tatarinov, I., Viglas, S., Beyer, K.S., Shanmugasundaram, J., Shekita, E.J., Zhang, C.: Storing and querying ordered XML using a relational database system. In: Proceedings of the 2002 ACM SIGMOD International Conference on Management of Data, Madison, Wisconsin, USA, 3–6 June 2002, pp. 204–215 (2002)
58. Tom, M.: A visual guide to the twisted web created by the Uber/Didi merger (2016). https://pitchbook.com/news/articles/a-visual-guide-to-the-twisted-web-created-by-the-uberdidi-merger
59. Tran, V.-D., Kato, H., Hu, Z.: BIRDS: programming view update strategies in datalog. PVLDB **13**(12), 2897–2900 (2020)
60. Tran, V.-D., Kato, H., Hu, Z.: A counterexample-guided debugger for non-recursive datalog. In: Oliveira, B.C.S. (ed.) APLAS 2020. LNCS, vol. 12470, pp. 323–342. Springer, Cham (2020). https://doi.org/10.1007/978-3-030-64437-6_17
61. Tran, V.-D., Kato, H., Hu, Z.: Programmable view update strategies on relations. PVLDB **13**(5), 726–739 (2020)
62. Tran, V.-D., Kato, H., Hu, Z.: Toward recursive view update strategies on relations. In: Ninth International Workshop on Bidirectional Transformations (BX 2021), June 2021
63. Ullman, J.D.: Information integration using logical views. In: Afrati, F., Kolaitis, P. (eds.) ICDT 1997. LNCS, vol. 1186, pp. 19–40. Springer, Heidelberg (1997). https://doi.org/10.1007/3-540-62222-5_34

# Contributed Papers

# Robust Cardinality Estimator by Non-autoregressive Model

Ryuichi Ito[✉], Chuan Xiao, and Makoto Onizuka

Osaka University, 1-1 Yamadaoka, Suita-shi, Osaka 565-0871, Japan
{ito.ryuichi,chuanx,onizuka}@ist.osaka-u.ac.jp

**Abstract.** In database systems, cardinality estimation is a fundamental technology that significantly impacts query performance. Recently, machine-learning techniques are employed for cardinality estimation, which learns dependencies among attributes. However, they have a problem that the estimation accuracy is unstable and the inference speed is slow. In this paper, we propose a stable and fast cardinality estimation method that learns dependencies among attributes by a non-autoregressive model and performs estimation in fewer steps and proper order according to a given query at the inference phase.

**Keywords:** Cardinality estimation · Machine learning

## 1 Introduction

As the use of data becomes more widespread, there are growing needs for faster database systems. Especially, the functionality of query optimizers is essential. The query optimizer consists of three main components: cost model, plan enumeration, and cardinality estimation. We focus on cardinality estimation, because Leis *et al.* [5] reported that cardinality estimation has the most significant influence on query performance among the three components. Cardinality estimation has been studied for many years [1,3,4,7,10,11], but practical database systems still prone to large estimation errors, up to $10^4 - 10^8\times$, which degrades the query performance [5]. Conventional methods for cardinality estimation maintain uniform value distributions as histograms for each attribute independently (*Uniformity, independence, and principle of inclusion assumption* [5]), and such cardinality estimation for queries on a single attribute is known to be sufficiently accurate. However, real-world value distributions among multiple attributes are not independent. The independence assumption causes major estimation errors in selection and join operations over multiple attributes. In recent years, cardinality estimation methods based on machine learning have been proposed [1,3,4,10,11]. Among them, methods of learning correlations among multiple attributes, which use autoregressive models [2,9] or sum-product

This paper is based on results obtained from a project, JPNP16007, commissioned by the New Energy and Industrial Technology Development Organization (NEDO).

G. Fletcher et al. (Eds.): SFDI 2021, CCIS 1457, pp. 55–61, 2022.
https://doi.org/10.1007/978-3-030-93849-9_3

networks [6], outperform conventional methods. The advantage of these methods is that they do not require any prior workload since they use only the database for training. However, there is a problem that the performance of these methods is not stable. For example, in the case of autoregressive models, their estimation accuracy largely depends on the order of attributes that needs to be fixed at the training phase.

In this paper, we propose a fast and stable cardinality estimation with a non-autoregressive model for learning correlations among multiple attributes. By utilizing the non-autoregressive model as a density estimator, we can capture a correlation among any combinations of attributes in multiple relations. Unlike autoregressive models, whose attributes order for inference is fixed at the training phase, non-autoregressive models learn all attributes in parallel and can infer conditional probabilities of each attribute in any order. This enables cardinality estimation to mitigate the performance instability due to the fixed order of attributes with autoregressive models. We evaluate our method using two real-world datasets and show that it can reduce Q-Error [5] to up to $1/35$ compared to PostgreSQL and be up to $2-3\times$ faster than the existing methods using the autoregressive model while stably achieving the comparable performance as its peak performance.

## 2    Formulation

In this chapter, we present a formulation for cardinality estimation using a density estimation model. Notice that cardinality is calculated from the joint probability satisfying a selection condition of a given query. Suppose that a relation $R$ consists of attributes $\boldsymbol{A} = \{A_1, ..., A_n\}$ and tuples $\{t_1, ..., t_m\}$, and operations of a selection condition are expressed with $\boldsymbol{\theta}(t) \rightarrow \{0, 1\}$. The cardinality $\mathcal{C}(\boldsymbol{\theta})$ is equal to the number of tuples that satisfy the selection condition $\boldsymbol{\theta}$ or the product of the total number of tuples and the probability of satisfying the selection condition:

$$\mathcal{C}(\boldsymbol{\theta}) = |\{t \in R \mid \boldsymbol{\theta}(t) = 1\}| \qquad (1)$$
$$= |R| \cdot P(\boldsymbol{\theta}(t) = 1) \qquad (2)$$

Each tuple $t$ consists of combinations of attribute values $\{a_1 \in A_1, ..., a_n \in A_n\}$. Thus, the cardinality $\mathcal{C}(\boldsymbol{\theta})$ can be defined as follows, where $P(A_1 = a_1, ..., A_n = a_n)$ is the joint probability:

$$\mathcal{C}(\boldsymbol{\theta}) = |R| \cdot \sum_{a_1 \in A_1} ... \sum_{a_n \in A_n} \boldsymbol{\theta}(a_1, ..., a_n) P(A_1 = a_1, ..., A_n = a_n) \qquad (3)$$

However, as can be seen from the above equation, it is difficult to deal with all the joint probabilities directly because they are numerous. Therefore, by assuming that each attribute is independent $P(A_1, ..., A_n) \approx \prod_n P(A_i)$, obtaining the cardinality by combining the histograms of multiple attributes is widely used.

However, this assumption is inappropriate for real-world data because there are usually correlations among attributes; so it is a major source of errors.

In this paper, we use the multiplication theorem, which is an exact equation, instead of the independence assumption. If $\boldsymbol{\theta}$ is conjunctive ($\boldsymbol{\theta} = \theta_1 \otimes \theta_2 \otimes ... \otimes \theta_n$, $\theta_i(a_i) \rightarrow \{0,1\}$), then cardinality $\mathcal{C}$ can be defined in the following form:

$$
\mathcal{C}(\boldsymbol{\theta}) = \begin{aligned} |R| \cdot \sum_{a_1 \in A_1} \theta_1(a_1)P(A_1 = a_1) \sum_{a_2 \in A_2} \theta_2(a_2)P(A_2 = a_2 \mid A_1) \\ ... \sum_{a_n \in A_n} \theta_n(a_n)P(A_n = a_n \mid A_{n-1}, ..., A_1) \end{aligned} \tag{4}
$$

## 3 Cardinality Estimation by Non-autoregressive Model

Our method employs non-autoregressive models[1] as its core to estimate cardinalities. Since our model is trained only on the distribution of data without requiring workloads, it can estimate the cardinalities of arbitrary queries with a single trained model. When the target database contains multiple relations, our method can estimate query cardinalities including join operations by regarding all relations as a universal relation with full outer join.

*Training.* We explain how we train a density estimator using a non-autoregressive model for cardinality estimation. The training is made on a tuple-by-tuple basis; each tuple with some masked attributes is used as input and the corresponding raw tuple is used as target. The losses of the masked attributes are used for updating parameters. By randomly selecting attributes for masking, the model can infer the probability distributions of arbitrary attributes with an arbitrary condition. Since domains are different among attributes, each attribute is embedded individually in the model.

*Inference.* As shown in Eq. 4, cardinality is computed by the product of the total number of tuples and the joint probability satisfying a selection condition; hence we estimate conditional probabilities in order by the model and derive a cardinality from them. We extend **Progressive Sampling** [11] for adapting it to non-autoregressive models. Although the multiplication theorem is an exact equation, Progressive Sampling uses the results of prior inference as input to its subsequent inference and the procedure depends on the Monte Carlo method. Therefore, the order of the expressions obtained based on the multiplication theorem affects the estimation performance. In particular, prior inference with low accuracy can harm the accuracy of the subsequent inference. In autoregressive models, the order of attributes needs to be fixed at the training phase. It is not easy to decide an appropriate order of attributes among $!|A|$ candidate orders. We must train an individual model from scratch for each candidate order of attributes. In contrast, non-autoregressive models can infer conditional probabilities of only needed attributes in any order. It can reduce inference costs.

---

[1] Any non-autoregressive model (e.g., Transformer-based model) can be used.

Therefore, by introducing a heuristic that inference becomes easier and more accurate in ascending order of the size of the domain, our method performs inference on attributes in ascending order of the size of their domain satisfying a given selection condition. This order is dynamically decided at runtime. The procedure is shown in Algorithm 1.

---

**Algorithm 1.** Cardinality Estimation by Non-Autoregressive Model

Input: $\mathcal{M}, \theta, A, N, |R|$
Output: $cardinality$
1: **procedure** ESTIMATE($\mathcal{M}, \theta, AttrsInPreds$))
2:    Initialize $inputs$ w/ nulls
3:    $prob \leftarrow 1.0$
4:    **foreach** $A_i \in AttrsInPreds$ **do**
5:       $dist_{A_i} \leftarrow \mathcal{M}(inputs)$                              ▷ Forward
6:       $dist'_{A_i} \leftarrow \{\theta_i(a_i)dist_{a_i} \mid a_i \in A_i\}$              ▷ Filtered by $\theta$
7:       $prob \leftarrow prob * \sum_{a_i \in A_i} dist'_{A_i}$
8:       $a_i \leftarrow$ SAMPLING($dist'_{A_i}$)
9:       $inputs[A_i] \leftarrow$ ENCODE($a_i$)
10:    **end for**
11:    **return** $prob$
12: **end procedure**
13:
14: $AttrsInPreds \leftarrow \{A \in A \mid A \text{ USED IN } \theta\}$
15: **foreach** $i \in N$ **do**                              ▷ Batched in practice
16:    $probs[i] \leftarrow$ ESTIMATE($\mathcal{M}, \theta,$
17:       SORTBYFILTEREDDOMAINSIZE($AttrsInPreds$))
18: **end for**
19: $selectivity \leftarrow$ MEAN($probs$)
20: $cardinality \leftarrow selectivity * |R|$
21: **return** $cardinality$

---

This algorithm, an extension of Progressive Sampling, takes a non-autoregressive model $\mathcal{M}$, a selection condition $\theta$, attributes $A$, sample size $N$, and the total number of tuples $|R|$. Because $\sum_{a_i \in A_i} P(a_i \mid \cdot)$ is always 1, there is no need to infer probability of attributes not included in the selection condition. Therefore, we extract only attributes included in the selection condition (1.14). Next, we infer conditional probability distributions ($dist_{A_i}$) and calculate a joint probability ($prob$) based on the multiplication theorem $N$ times (1.15–18). At this time, the attributes are given in ascending order of the size of the domain satisfying the selection condition (1.17). We apply the selection condition to inferred probability distribution $dist_{A_i}$ and multiply the sum of them and $prob$ (1.7). In addition, we sample a value $a_i$ of $A_i$ based on the inferred probability distribution for the next inference (1.8). By repeating these operations on all the attributes included in the selection condition, we obtain the estimated selectivity (1.11). Finally, we return the product of the mean of $N$ predicted selectivities $probs$ and the total number of tuples $|R|$ as the estimated cardinality (1.19–21).

## 4   Experiments

We evaluate our method and discuss it based on the results. We implement our method using multilayer perceptron (MLP) and Transformer (Trm) [9][2], as non-autoregressive models. Note that *Prop.* order represents the ascending attribute

---

[2] Only the decoder.

order based on the size of the attribute domain satisfying the selection condition. We use two benchmarks on real-world datasets for evaluation, DMV [8] and Join Order Benchmark [5] (JOB-light [4]). DMV consists of queries with multiple equal and range conditions. JOB-light consists of queries with multiple equal conditions, range conditions, and join operations. For comparison, we use the cardinality estimator used in PostgreSQL 11.8 and Naru [11]/NeuroCard [10], which are state-of-the-art cardinality estimators using autoregressive models. NeuroCard is an extended method of Naru for multiple relations. Both methods are mainly implemented with MADE, but secondarily Transformer as well. For the attribute order, which is a hyperparameter of Naru/NeuroCard, we use *Natural* order (left-to-right order in each table schema) that is reported to be the best [11] and *Reverse* order that is the reverse order of *Natural* for comparing variances. The execution environment, hyperparameters, sampled tuples for training, etc., are set to be as equal as possible between Naru/NeuroCard and our method. The sample size for the Monte Carlo method was fixed to 1000 for all the methods. In these benchmarks, we have confirmed in preliminary experiments that there is almost no change in performance when the sample size is increased from 1000. We use Q-Error, which indicates how many times an estimated cardinality differs from a true value, as an evaluation metric. Q-Error is always greater than or equal to 1 and smaller values indicate better performance.

We show the overall evaluation results with Q-Error (median, 90th percentile, 95th percentile, 99th percentile, max, and the coefficient of the variation of the samples) and the response time of each method.

**Table 1.** Q-Error and mean latency (ms) on *DMV*

| Method (Impl. · Order) | Median | 90th | 95th | 99th | Max | CV | Latency |
|---|---|---|---|---|---|---|---|
| PostgreSQL | 1.27 | 2.22 | 57:9 | $1.4 \cdot 10^3$ | $7.6 \cdot 10^4$ | – | 2.93 |
| Naru (MADE · Natural) | 1.04 | 1.19 | 1.32 | 2.00 | 8.00 | 0.43 | 10.3 |
| Naru (MADE · Reverse) | 60.8 | $5.1 \cdot 10^2$ | $7.5 \cdot 10^2$ | $2.0 \cdot 10^3$ | $4.2 \cdot 10^5$ | 3.01 | 10.4 |
| Naru (Trm · Natural) | 1.09 | 2.63 | 5.34 | $9.7 \cdot 10^5$ | $9.9 \cdot 10^5$ | 0.45 | 99.5 |
| Naru (Trm · Reverse) | 1.18 | 10.8 | $2.0 \cdot 10^2$ | $1.0 \cdot 10^3$ | $2.1 \cdot 10^4$ | 0.78 | $1.0 \cdot 10^3$ |
| Ours (MLP · Prop.) | 1.05 | 1.30 | 1.56 | 2.60 | 49.0 | 0.10 | 5.76 |
| Ours (Trm · Prop.) | 1.04 | 1.22 | 1.41 | 2.33 | 49.0 | 0.10 | 54.3 |

Table 1 shows that the performance of Naru varies more than 10 times depending on the order of attributes for both MADE and Transformer implementation. Namely, the performance of Naru cannot be stable unless an appropriate order of attributes is selected. In contrast, our method stably achieves the comparable performance as Naru (MADE, Natural), which is the best setting of Naru in the experiment. Besides, Table 2, whose benchmark contains multiple natural joins, shows that our method is stable and outperforms many others. However, in rare cases, we observe that our method extremely underestimates

**Table 2.** Q-Error and mean latency (ms) on *JOB-light*

| Method (Impl. Order) | Median | 90th | 95th | 99th | Max | CV | Latency |
|---|---|---|---|---|---|---|---|
| PostgreSQL | 7.44 | $1.6 \cdot 10^2$ | $8.4 \cdot 10^2$ | $3.0 \cdot 10^3$ | $3.5 \cdot 10^3$ | – | 3.88 |
| NeuroCard (MADE · Natural) | 1.79 | 9.00 | 19.5 | 36.5 | 43.2 | 5.07 | $1.6 \cdot 10^2$ |
| NeuroCard (MADE · Reverse) | 6.04 | 72.3 | $1.2 \cdot 10^2$ | $3.0 \cdot 10^2$ | $4.0 \cdot 10^2$ | 7.18 | $1.6 \cdot 10^2$ |
| Ours (MLP · Prop.) | 3.14 | 25.0 | 60.5 | $7.7 \cdot 10^2$ | $2.2 \cdot 10^3$ | $7.3 \cdot 10^{-2}$ | 46.5 |
| Ours (Trm · Prop.) | 2.41 | 11.8 | 16.8 | $4.1 \cdot 10^3$ | $1.3 \cdot 10^4$ | 4.39 | $1.9 \cdot 10^3$ |

cardinalities due to some attribute values with very low frequency and results in large Q-Error. This behavior is not observed in NeuroCard. One of the reasons for the difference is that our model overfits to the samples. Therefore, in the case of training only on the sampled tuples, we should employ methods to suppress the overfitting for solving the problem. Overall, the latency results confirmed that our method is $2-3\times$ faster than autoregressive-based methods of the same scale.

## 5    Conclusion and Future Work

This paper proposed a stable and fast cardinality estimation using a non-autoregressive model. By learning a distribution of data and applying a selection condition at the inference phase, our method achieves stable cardinality estimation. We evaluate our method using Q-Error as a metric for real-world datasets. The results show that our method achieves $2-3\times$ faster than the existing methods using autoregressive models while achieving the comparable performance as the peak performance of the existing methods.

In the future, we plan to tune hyperparameters and automate its procedure. In addition, we plan to use more complex benchmarks which contain updates and to make evaluations of query optimizers using estimated cardinalities by our method.

## References

1. Chow, C., Liu, C.: Approximating discrete probability distributions with dependence trees. IEEE Trans. Inf. Theory **14**(3), 462–467 (1968)
2. Germain, M., Gregor, K., Murray, I., Larochelle, H.: MADE: masked autoencoder for distribution estimation. In: PMLR (2015)
3. Hilprecht, B., Schmidt, A., Kulessa, M., Molina, A., Kersting, K., Binnig, C.: DeepDB: Learn from Data, not from Queries! VLDB (2019)
4. Kipf, A., Kipf, T., Radke, B., Leis, V., Boncz, P., Kemper, A.: Learned cardinalities: estimating correlated joins with deep learning. In: CIDR (2019)

5. Leis, V., Gubichev, A., Mirchev, A., Boncz, P., Kemper, A., Neumann, T.: How good are query optimizers, really? VLDB **9**(3), 204–215 (2015)
6. Poon, H., Domingos, P.: Sum-product networks: a new deep architecture. In: AAAI (2017)
7. Poosala, V., Ioannidis, Y.E., Haas, P.J., Shekita, E.J.: Improved histograms for selectivity estimation of range predicates. In: ICDE (1996)
8. State of New York: Vehicle, Snowmobile, and Boat Registrations. https://catalog. data.gov/dataset/vehicle-snowmobile-and-boat-registrations (2019)
9. Vaswani, A., et al.: Attention Is All You Need. In: NIPS (2017)
10. Yang, Z., et al.: NeuroCard: One Cardinality Estimator for All Tables. VLDB (2021)
11. Yang, Z., et al.: Deep Unsupervised Cardinality Estimation. VLDB (2019)

# Conflict Resolution for Data Updates by Multiple Bidirectional Transformations

Mikiya Habu[1] and Soichiro Hidaka[2]($\boxtimes$) (iD)

[1] FUJIFILM System Services Corp., Tokyo, Japan
[2] Hosei University, Tokyo, Japan
habutaso@gmail.com, hidaka@hosei.ac.jp

**Abstract.** Bidirectional transformation (BX) is a robust mechanism to propagate changes of data across the transformation while maintaining consistency between two or more data sources. Recently, systems that coordinate multiple BX have been proposed. However, conflicts when multiple BX update the same source have not been well studied yet. In this paper, we propose a new conflict resolution method for BX based on an algorithm of Operational Transformation (OT). We apply the algorithm of OT to the backward transformation of the tree duplication primitive Dup of an existing BX language X proposed by Hu et al. X had been shown, by embedding into Inv, which is capable of maintaining structured documents such as XML through BX, to allow more flexible operations by Inv's bidirectionality-satisfying operations. Using this mechanism, we propose a new backward transformation function *mput* that can resolve conflicts between updates on two views. Our implementation of *mput* can simultaneously satisfy the properties GETPUTGET and PUTGETPUT, which are the round-tripping properties inherited from X, and TP1, which is a confluence property inherited from OT for server-client environment.

**Keywords:** Bidirectional transformation · Operational transformation · Conflict resolution

## 1 Introduction

Bidirectional transformation (BX) is a mechanism to propagate changes across two or more information sources through BX rules, keeping consistencies between them. Compared to preparing two transformations for each direction and guaranteeing the consistency of the two transformations by the programmer, BX requires only one program. The consistency is automatically guaranteed by the well-behavedness properties that BX satisfies. Although most of the previous work on BX focus on those among two information sources, maintenance of consistency by combining multiple BX, as well as multi-directional transformations as a general mechanism to maintain consistency among three or more information sources are more considered recently [3]. For example, POET [9] uses

© Springer Nature Switzerland AG 2022
G. Fletcher et al. (Eds.): SFDI 2021, CCIS 1457, pp. 62–75, 2022.
https://doi.org/10.1007/978-3-030-93849-9_4

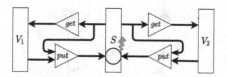

**Fig. 1.** BX in which *puts* face each other and the point of conflicts

multiple BX to allow privacy-aware data access defined for each edge (computing entity architecturally located closer to the end-devices) in edge computing. The forward transformation *get* in POET can extract views for data accessed by multiple edges. However, the backward transformation *put* may cause conflicts because of multiple edges accessing identical sources $S$. Figure 1 shows such a situation in which two BX are combined so that *put* transformations for two views $V_1$ and $V_2$ face with each other. Suppose that updates on two views are being propagated via two *puts* to their common source at the same time. Such conflicts cannot be resolved by existing (well-behavedness) properties of BX, as the authors of POET addressed in their future work [9]. We could resolve the conflict by prioritizing one of these updates, while discarding the other update. Here, the case degenerates to that in which there are only one pair of source and view. However, updates by one user are abandoned no matter how important the updates are. Basically, BX maintain consistency in a robust manner for each combination of source and view, whereas situations in which there are multiple sources or views are not so well-studied, especially in the presence of conflicts.

In this paper, we cope with this situation by providing data update mechanism so that no user is sacrificed, taking each user's update into account. Existing techniques of operational transformations (OT) are used to merge such concurrent updates. The rest of this paper is organized as follows. Sections 2 and 3 briefly introduce the basic notions of BX and OT. Section 4 touches on previous work and conferences focusing on multiple BX. Section 5 recaps OT on tree data structures that are necessary to resolve conflicts in structured documents. We borrow this technique and combine with BX. In Sect. 6, we define a novel BX called *mget* and *mput* using existing BX language X [7,8] for structured documents, and one of its combinators Dup. Section 7 describes our implementation strategy using the Inv language in which the X language is embedded. Section 8 reports the results from our validation on whether our implementation of *mput* exhibits a correct behavior. Section 9 concludes our paper and mentions the remaining issues and outlook. Our implementation as well as the test code for our evaluation can be found at https://github.com/habutaso/x-inv-xeditor-dev.

## 2   Bidirectional Transformations

In BX, two information sources are denoted by source $S$, and view $V$, respectively. For a given BX $f$ between $S$ and $V$, the operation to extract data of interest from $s \in S$ and construct $v \in V$ is called *get* and denoted by $\phi_f : S \to V$. The other operation, given a view $v \in V$ that contains certain updates, that reflects

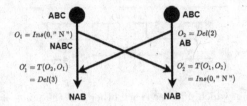

**Fig. 2.** A transformation process in operational transformations

such updates to $s \in S$ is called backward transformation *put* and is denoted by $\lhd_f : V \times S \to S$. The operator $\lhd$ is a function that takes one element from each $V$ and $S$, and returns an element in $S$. The subscript $f$ represents one of the combinators of X language [7,8] recapped in Sect. 6 that provides transformation rules for forward and backward directions. BX typically satisfy the following two properties.

$$s \lhd (\phi s) = s \qquad (\text{GetPut})$$
$$\phi (s \lhd v) = v \qquad (\text{PutGet})$$

GetPut says that the source remains the same state after the backward transformation if no change has been made on the extracted view, whereas PutGet says that the backward transformation can convey all the information in the view to the source. A pair of BX has *well-behavedness* property iff it satisfies both GetPut and PutGet. This property exhibits how well a pair of bidirectional transformations behaves. The main advantage of BX is to be able to propagate changes bidirectionally while maintaining such properties.

Conventionally, *put* assumes only one view. We have to introduce a new mechanism if we consider two views. In Sect. 6.5, we describe BX functions *mget* and *mput* required for two views.

## 3   Conflict Resolution Algorithms

Operational transformation (OT) [6] is one of the algorithms that have been proposed for conflict resolution. OT is a technology to support collaborative work on documents and data structures in collaborative editing software. Currently this technology is adopted to G Suite [2] and Etherpad[1], among other systems, to achieve consistency maintenance under concurrency. Figure 2 illustrates the process of transformation of operations in collaborative document editing software. In particular, it shows the transformation property called TP1 or $C_1$ where the result of editing for each editor becomes equivalent by incorporating operations on the other sides with appropriate adustments (i.e., transformations). Each of the black circles on the upper part of the figure denotes a user who shares an

---
[1] https://etherpad.org/.

identical string "ABC" in the editing session. $O_1$ and $O_2$ denote editing operations. Suppose the user on the left side executed *Ins* to insert the character "N" at position 0, while the user on the right side executed *Del* operation to delete the second character, simultaneously. At this point, the left user holds the string "NABC", while the user on the right holds "AB", causing conflict in the session. Then the two operations are exchanged, and each recipient invokes the transformation $T$ which calculates an editing operation by which the final strings agree, by taking their own operation and received operation as arguments. Each user holds the string "NAB" in the end, thus the consistency of the shared document is maintained. OT are often deployed in server-client architecture, where each client stores their editing operations in a buffer and periodically executes the possibly transformed editing operations. Thanks to this mechanism, not only the conflicts among operations that occur simultaneously, but also those that occur at different moments.

In this paper, we aim at applying the OT algorithm to resolve conflicts when two *put* transformations try to propagate changes on their views to their shared source.

## 4   Related Work

Diskin proposed the lane mechanism [4] for conflict resolution in BX. Although his work mainly focuses on model synchronization tools, the synchronization mechanism uses BX that worth investigation. When two BX try to update their shared views through their *get* transformations, they do not update their views directly, but they keep their abstract data as replicas the results obtained by *get*s, and relate them with an abstract algorithm based on category theory. On the contrary, we do not require such replicas, but apply an OT algorithm on propagating updates without changing the data manipulated inside component BX. Besides, we enable data updates in which updates on multiple views are considered. This has advantages of avoiding data loss caused by conflicts. In the Dagstuhl Seminar [3] on Multidirectional Transformations and Synchronisations in 2018, the issue of conflict resolution was discussed as one of the open problems. It was considered possible as a simple conflict resolution strategy to keep timestamps of executions of different BX, so that the transformations are applied according to the order of the timestamps. However, the idea of conflict resolution using OT was not proposed in the seminar report [3].

## 5   OT on Trees

Early OT resolved conflicts on texts, as we introduced in Sect. 3. However, since we study BX on structured documents like XML, we need a mechanism to transform operations on trees. We borrow the methods proposed by Sinchuk et al. [11] for this purpose. Sinchuk et al. represent trees by the following Rose Trees.

$$Tree\ \tau ::= \text{Node}\ \tau\ [Tree\ \tau]$$

```
list_it ot (TreeRemove n1 l1) (TreeInsert n2 l2) flag
   | n1 + len1 < n2 = op1 : []
   | n2 < n1 = (TreeRemove (n1 + len2) l1) : []
   | otherwise =
       let ml = ins (n2 - n1) l2 l1 in
       if isJust ml then
       let Just l' = ml in (TreeRemove n1 l') : []
       else op1 : []
   where op1 = TreeRemove n1 l1
         len1 = length l1; len2 = length l2
```

**Listing 1.** The conflict resolution program on the side who issued TreeInsert

Since the Node constructor holds node labels of type $\tau$ and a list of nodes of its own type as its subtree, the OT rules are defined for both trees and lists. For a given tree, the following four operations can be transformed.

1. TreeInsert $i$ $t$: inserts tree $t$ as $i$-th subtree.
2. TreeRemove $i$ $t$: deletes the $i$-th subtree if it is equal to $t$.
3. EditLabel $tc$: replaces the root label using operation $tc$. The operation $tc$ depends on the label type $\tau$, so it is separately determined by developers.
4. OpenRoot $i$ $c$: applies operation $c$ on the $i$-th subtree.

The property illustrated by Fig. 2 is called $C_1$ in Sinchuk et al.'s paper [11], meaning that for the transformation function $T$ to resolve conflicts of operations $O_1$ and $O_2$ on their issuer's site, the following equation always holds.

$$O_1 \circ T(O_2, O_1) \equiv O_2 \circ T(O_1, O_2) \tag{$C_1$}$$

Note that the composition $\circ$ means the operations are applied in the order from left to right. The first and second argument of $T$ corresponds to the remote (counterpart's) and local operations, respectively. It was mechanically proved using Coq [1] that $C_1$ is satisfied by Sinchuk et al.'s OT on trees. There exists another desirable property $C_2$ for OT. However, $C_2$ assumes a rather rare situation, compared with more common environment where existence of servers can be assumed. With this reason, $C_2$ is not considered mandatory and omitted in their work.

In this paper, we use String as $\tau$ and let the operation $tc$ replace the current value with the input string. When we are to transform the conflicting EditLabel operations, Sinchuk et al. set priorities using boolean flags, such that when we compute $T$ (EditLabel $tc_1$, EditLabel $tc_2$), assuming true for editor 1 (local), false for editor 2 (remote), operations of $tc_2$ takes precedence. For further details on how two operations are transformed by $T$, see their implementation [11].

In this section, we take as an example the case in which the two copies of the tree Node "r" [Node "a" [ ], Node "b" [ ], Node "c" [ ]] of type *Tree* String are manipulated. Let the two editors issue operations TreeInsert 2 [Node "d" [ ]] and TreeRemove 0 [Node "a" [ ]], respectively, to cause conflict. Sinchuk et al. resolved such conflicts by the function list_it that corresponds to $T$.

```
list_it ot (TreeInsert n1 l1) (TreeRemove n2 l2) _
    | n1 < n2 = op1 : []
    | (n2 + len2) < n1 = (TreeInsert (n1 - len2) l1) : []
    | otherwise = []
    where op1 = TreeInsert n1 l1
          len2 = length l2
```

**Listing 2.** The conflict resolution program on side who issued TreeRemove

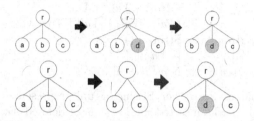

**Fig. 3.** Tracing transformed operations: upper: former lower: latter

Listings 1 and 2 are its Haskell implementations. They are manually adapted from the Coq code in Sinchuk et al.'s paper [11] and the Haskell code generated from the Coq code. The algorithm basically shifts the positions of operations according to the relative positions of given conflicting operations. The case analysis in the transformations of operated positions are implemented using Haskell guards. We apply $C_1$-compliant operations based on this algorithm.

In our example, the conditions on the matching relative positions are colored in red inside Listings 1 and 2. After transformation, the former applies TreeInsert 2 [Node "d"  [ ]] followed by TreeRemove 0 [Node "a"  [ ]], while the latter applies TreeRemove 0 [Node "a"  [ ]] followed by TreeInsert 1 [Node "d" [ ]]. Figure 3 traces how the trees on both sides are manipulated.

## 6   X/Inv Language and Its Extension

The conflict resolution in the OT inspects the operations themselves and positions on which the operations are performed. Therefore, such information should be treated in BX as well. The language X by Hu et al. [7] is a BX language in which not only the data (states) themselves but also editing operations performed on these data can be propagated bidirectionally. Thanks to such framework, BX considering insertion, deletion and duplication are possible. The combinator Dup is introduced as a special atomic combinator to support data duplication. Forward transformation of Dup is capable of data duplication, while the backward transformation can merge the duplicates. Such mechanism can be considered to support more flexible manipulation of data compared to general BX. In this paper, we apply the Dup of X language to implement backward transformation that allows two views as inputs.

## 6.1   Data Structure

First of all, we recap the definition of the data type *Val*, as a data structure to transform editing operations bidirectionally, that also embeds the information of insertion, deletion and label modifications, in the data on which the updates are propagated.

$$
\begin{aligned}
Val &\;::=\; Atom \\
&\;\;\mid\; *Atom \mid Val^+ \mid Val^- \\
&\;\;\mid\; (Val \times Val) \mid [Val] \mid Tree\ Val \\
Atom &\;::=\; String \mid () \\
[a] &\;::=\; [\,] \mid a : [a] \\
Tree\ a &\;::=\; N\ a\ [Tree\ a]
\end{aligned}
$$

Atomic data resources are either String or unit represented by (). $*$ marks updated *Atoms*. $+$ marks inserted, and $-$ marks deleted *Vals*. They are also called tags. Colons (:) denote cons of lists. The notation is similar to that of Haskell, and "A":"B":[] can be rewritten into ["A", "B"]. *Tree* has constructor N, holding node labels of type $a$ and a list of their children of type *Tree a*. *Val* is a simple tree data structure, omitting several XML constructs such as attributes.

## 6.2   Dup Combinator

In this subsection, we describe the characteristic atomic operator Dup in the language X. Let $c$ a source (concrete view). Then $\phi_{\mathrm{Dup}}$ is defined as follows.

$$
\phi_{\mathrm{Dup}}\ c = N\ \texttt{"Dup"}\ [c,\ c]
$$

Here, N denotes the node constructor of trees, having the root node labeled "Dup" as well as two identical children $c$. Meanwhile, given $a_1, a_2$ as the view (abstract view), $\triangleleft_{\mathrm{Dup}}$ is defined as follows.

$$
c \;\triangleleft_{\mathrm{Dup}}\; (\text{Node "Dup"}\ [a_1,\ a_2]) =
\begin{cases}
a_2 & (a_1 = c) \\
a_1 & (a_2 = c) \\
a_1 & (\text{otherwise})
\end{cases}
$$

The user is assumed to perform only one editing operation at a time. When either of $a_1, a_2$ is updated, then that update takes precedence and is propagated to the source. If both are updated, then $a_1$ is always chosen.

## 6.3   The Language Inv and Embedding of X

The language Inv [7,10] updates structured documents bidirectionally. It defines binary relations on *Val* described in Sect. 6.1. The atomic data operations in Inv can be combined. The reason why we use Inv is that combinators in X can be given bidirectionality by expressing them in $Inv^2$. We introduce a new operator in X/Inv in this paper to extend Dup operator.

---

[2] Compared to direct bidirectionalization [7], convenience can be improved because Inv-specific language constructs such as fixpoint combinators become available.

**Syntax.** Inv has the following syntax. See [10] for their detailed semantics.

$$\begin{aligned}
\text{Inv} ::=\ & \text{Inv}^{\smile} \mid nil \mid cons \mid node \mid P? \mid V \\
& \mid\ \delta \mid dupNil \mid dupStr\ \text{String} \\
& \mid\ \text{Inv}; \text{Inv} \mid id \mid \text{Inv} \cup \text{Inv} \\
& \mid\ \text{Inv} \times \text{Inv} \mid assocr \mid assocl \mid swap \\
& \mid\ \mu(V : \text{Inv}) \mid prim(f, g) \mid resC
\end{aligned}$$

The non terminal $P$ denotes predicates on $Val$, while $V$ denotes the set of variable names. The construct $resC$ is the special function newly introduced by us only used to resolve conflicts on two views. Although $resC$ is used for the backward transformation of the extended Dup, the reversed semantics is also defined. Inv is basically a point-free language, so that data can be processed through combinations of functions. In this paper, we only explain Inv functions needed to extend Dup.

Figure 4 shows the binary relations on $Val$ denoted through semantic function $[\![\_]\!]_\eta$. The environment $\eta$ is the binary relations on $Val$ between source and view. In Inv, the binary relation is extracted from the environment $\eta$ and the forward/reverse transformations are conducted under that environment. The mapping $\eta$ is used because, in Inv, the fixpoint operator $\mu(V : \text{Inv})$ is introduced so that various functions are composed in a point-free style without adding new primitive functions. The fixpoint operator injects entries to the environment.

The underscore corresponds to Inv language constructs. $^{\smile}$ denotes reverse transformation in Inv and performs reverse operation of given functions. For example, $nil^{\smile}$, the reverse of $nil$, takes empty list as input and returns unit () as the output. $cons^{\smile}$ decomposes non-empty lists into its first element and the rest. $node^{\smile}$ extracts from N its label $a$ and the list of its children $x$ as a tuple. $id^{\smile}$ and $swap^{\smile}$ perform the same operation as their forward direction. $\delta$ duplicates atomic values only. $resC$, our new operator, takes two inputs. $a$ corresponds to the source, while $[a, a]$ corresponds the two views encoded as list with the two values as its elements. When the source $a$ and two views are identical, then there is no need to update the source, so $a$ is returned as is. When only one of the views of source $a$ contains a tag to encode editing operations on the view, then either pattern $[a', a]$ or $[a, a']$ matches and the editing operations are propagated to produce $a'$ as the output. Further, when both of the two views are associated with editing tags, then $a'''$ in which conflicts caused by $a'$ and $a''$ are resolved by OT are produced to reflect the edits to the source. In its reverse direction, the result is isomorphic to that of the forward direction of $\delta$.

**Dup in Terms of Inv.** The combinator $x$ in X that is embedded in Inv is denoted by $\lceil x \rceil$. Dup construct to duplicate its inputs of type $a$ is expressed as follows.

$$\begin{aligned}
\lceil \text{Dup} \rceil &= dup_a; (id \times (dup_a; futatsu; mkRoot)) \\
\text{where}\ futatsu &= (id \times (dupNil; cons)); cons \\
mkRoot &= dupstr\ \text{"Dup"}; swap; node
\end{aligned}$$

$$[nil]_\eta \; () \quad = [\,]$$
$$[cons]_\eta \; (a,x) = a : x$$
$$[node]_\eta \; (a,x) = N \; ax$$
$$[id]_\eta \; a \qquad = a$$
$$[swap]_\eta \; (a,b) = (b,a)$$
$$[dupNil]_\eta \; a \quad = (a,[\,])$$
$$[dupStr \; s]_\eta \; a \; = (a,s)$$
$$[\delta]_\eta \; n \qquad = (n,n)$$
$$[\delta^\backprime]_\eta \; (n,n) \quad = n,$$
$$n \text{ not tagged or tagged by } ^{+,-}$$
$$[\delta^\backprime]_\eta \; (*n, *n) = *n$$

$$[\delta^\backprime]_\eta \; (*n, m) \qquad = *n, \; m \text{ not tagged}$$
$$[\delta^\backprime]_\eta \; (n, *m) \qquad = *m, \; n \text{ not tagged}$$
$$[f;g]_\eta \qquad = [f]_\eta ; [g]_\eta.$$
$$[f \times g]_\eta \qquad = ([f]_\eta \times [g]_\eta)$$
$$[resC^\backprime]_\eta \; a \qquad = [a,a]$$
$$[resC]_\eta \; (a,[a,a]) \quad = a$$
$$[resC]_\eta \; (a,[a',a]) \quad = a'$$
$$[resC]_\eta \; (a,[a,a']) \quad = a'$$
$$[resC]_\eta \; (a,[a',a'']) \; = a'''$$
$$a \text{ not tagged}$$
$$a' \text{ and } a'' \text{ tagged by } ^{+,-,*}$$
$$a''' \text{ retains resolved conflict between } a' \text{ and } a''$$

**Fig. 4.** Inv constructs

$dup_a$ is called twice, once for caching input, and the other to perform actual duplication. The operation *futatsu*; *mkRoot* will produce a tree with "Dup" as its label from a pair of trees ("futatsu" means "two of" in Japanese).

### 6.4   Bidirectionality

In BX, it is important to satisfy PUTGET and GETPUT. However, in language X, PUTGET cannot be satisfied. For example, $c \; \triangleleft_{\text{Dup}} (N \; "Dup" \; [a, *b])$ returns $b$ (* denotes nodes with its labels updated). Further, when $b$ is applied to $\phi_{\text{Dup}}$, $N \; "Dup" \; [b,b]$ is returned. Therefore, it is not well-behaved. Instead, the following properties in which GETPUT and PUTGET are relaxed are satisfied.

$$\phi \, (s \triangleleft v) = v \qquad \qquad \text{(GETPUTGET)}$$
$$s' \; \triangleleft \; (\phi \, s') = s' \qquad \qquad \text{(PUTGETPUT)}$$
$$\text{where } v = \phi \, s, \; s' = s \triangleleft v$$

In GETPUTGET , $v$ is produced by the forward transformation, $s$ is updated through $v$, after which the view is immediately computed to obtain $v$. In PUTGETPUT , when $s'$ is the recently updated source, a view is produced from the source, after which the backward transformation is immediately conducted to produce $s'$ again. If the two properties are satisfied, the users have only to perform one *put* and one *get* when they perform updates.

Based on the explanations so far, Dup is capable of duplicating views while being a BX combinator.

### 6.5   *mget* and *mput*

Since *put* assumes in general only one view, *put* alone cannot consider multiple views as described in Sect. 1. Therefore, we define the new functions *mget*, *mput* to cope with such situation. First, let $x.get : S \to V$ and $x.put : S \times V \to$

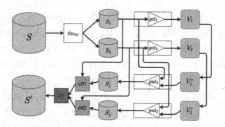

**Fig. 5.** BX to resolve conflicts in two views

$S$ denote *get* and *put* component, respectively, of BX $x$. Next, we provide a BX $x_1 \bigtriangleup x_2$ to deal with two views. We define its forward transformation as $(x_1 \bigtriangleup x_2).get \; s \triangleq (x_1.get \; s, x_2.get \; s)$. For cooperating $x_1$ and $x_2$, we uniquely determine the application of its *get*. Based on these operations, we define the following, where the operator $\vartriangleleft'$ is defined in the next section.

$$mget(s) = (x_1 \bigtriangleup x_2).get \; s$$
$$mput(s, v_1, v_2) = (x_1 \bigtriangleup x_2).put \; (s, v_1, v_2)$$
$$\triangleq \; \vartriangleleft' (s, s_1, s_2)$$
$$\text{where } s_1 = x_1.put \; (s, v_1)$$
$$s_2 = x_2.put \; (s, v_2)$$

Figure 5 shows the data propagation by *mget* and *mput*. The upper half corresponds to *mget*, while the lower half corresponds to *mput*. *mget* duplicates $s$ to produce two views. For each view, one *get* is executed to duplicate the source. In *mput*, two views generated by the duplication in *mget* are merged through the two *puts*. Peculiar to *mput* function, when multiple updates exist in the input views, the conflict should always be resolved.

## 7 Design of Our Proposal

The design of our proposal in our paper consists of two main ideas. The first one is the data transformation in *mput* that introduce new concept in BX to reflect changes on two views to the source. The other one is how to formalize the first one. In this section, we first explain a concrete process until the propagated updates reach the source, and then, we explain its more formal treatment. In the implementations that we used as a reference, the Haskell program of X implemented by Hu et al., as well as the Haskell program produced by the Coq program implemented by Sinchuk et al. as OT on trees.

### 7.1 Data Flows

Here we describe the design of data flow in the BX as an extension of X language. The view is assumed to be a client accessing sources. First, two clients access the common source and produce their views through their *gets*. In this paper, it is assumed that an identical datum is provided, for all clients, as their views before updates. Next, each client updates her/his view. Only one update primitive for each client is assumed. That is, only one operation is allowed for each client. The updated view is processed by $mput$. In $mput$, first $put_1$ and $put_2$ produce, respectively, $S_1'$ and $S_2'$ from views $V_1'$ and $V_2'$. Next, the differences representing updates are extracted from $S$, $S_1'$, and $S_2'$ Based on these differences, the conflicts are resolved through OT. Finally, the updates by $mput$ are completed by applying the new operation produced by the OT to the source. For each $put$, all the contents in the views are propagated, so it can be seen from the user's point of view a BX. Note that, in $mput$, $Val$ itself allows storage of editing operations, and editing operations are extracted from $Val$. Therefore, it can also be considered (internally) that a delta-based BX [5] is used. The OT, in general requires, for each client, corresponding applications of operations when operations after transformation is applied. However, in our proposal, since the target of application of the operation is only one, we need to choose one operation from two operations of both clients. Actually we do not have to care which to choose, because, as long as $C_1$ is satisfied, the final output on both clients are identical. In addition, we assume for $mget$, in this paper, $V_1$ and $V_2$ produced by $get_1$ and $get_2$, respectively, produce the identical views. We use this assumption to reuse $\phi$ in X, not to newly implement $mget$.

## 7.2   XEditor

In Hu et al.'s method [7], an editor to interactively edit structured documents is proposed. The source document with no internal dependency, and the mapping from the source to the view can be automatically generated. Editing operations are fed to the view, while its updates are propagated to the source through $put$. In this paper, we use the editing operations used in XEditor to implement OT within BX.

$$Command\ a ::= \text{Insert } Path\ a \mid \text{Delete } Path\ a$$
$$\mid\ \text{EditLabel } Path\ a \mid \text{Stay}$$
$$Path\ ::= [\text{Int}]$$

The editing operations are defined as follows. In this paper, $a$ is assumed to correspond to $Val$. $Path$ is represented by a list of non-negative integers $[i_1, i_2, \ldots, i_n]$. It identifies the subtree reached by first visiting the $i_1$-th child of the root, followed by visiting of $i_2$-th child, and so on, until $n$ traversals have been performed.

The semantics of the $Command$ is described for each operation as follows.

- Insert $Path\ a$: Insert $a$ at position $Path$.
- Delete $Path\ a$: Delete subtree at position $Path$. $a$ is present to align with the corresponding implementation of OT, so the value of $a$ is ignored. We used "___" (three underscores) as its dummy filler.
- EditLabel $Path\ a$: Update the label at position $Path$ to $a$.
- Stay: Do nothing.

Since the position $Path$ are explicitly specified for these editing operations, they can be associated with OT OpenRoot. Moreover, since each editing operation has similar semantics, we can transform $Command\ a$ using OT.

## 7.3   Design of *mput* in X

We apply the approach used in Dup described in Sect. 6.3 to design *mput*. In Sect. 6.3, we mention the new language construct *resC* in X/Inv introduced by us.

First, the original [8] embedding $\lceil Dup \rceil^{\vee}$ is expressed as follows ("hitotsu" means "one of" in Japanese).

$$\lceil Dup \rceil^{\vee} = (id \times (rmRoot; hitotsu; dup_a{}^{\vee})); dup_a{}^{\vee}$$
$$\text{where } rmRoot = node^{\vee}; swap; (dupStr \text{ "Dup"})^{\vee}$$
$$hitotsu = cons^{\vee}; (id \times (cons^{\vee}; dupNil^{\vee}))$$

Figure 6 depicts the processing flow of $\lceil Dup \rceil^{\vee}$. The black triangle denotes the source, the green and purple triangles denote conflicting subtrees, while the orange triangle denotes the subtree with the conflicts (insufficiently) resolved. Two $dup_a{}^{\vee}$ are performed in $\lceil Dup \rceil^{\vee}$. The first to merge the children of the node labeled "Dup", the second to merge the *Val* of the source of duplication with the *Val* of the result of the first merge. This form of merging is not useful for our method. This is because, when we compute the difference between the updated view and original view, the source and the two views have to be processed within one function. So we define the new embedding $\lceil Dup' \rceil^{\vee}$ that uses *resC* as follows. Figure 7 traces its transformation process. The coloring scheme in the figure is identical to that of Fig. 6, with further coloring is applied to compare the two figures.

$$\lceil Dup' \rceil^{\vee} = (id \times rmRoot); resC$$
$$\text{where } rmRoot = node^{\vee}; swap; (dupStr \text{ "Dup"})^{\vee}$$

The original embedding $\lceil Dup \rceil^{\vee}$ executed *hitotsu* after *rmRoot*, followed by merging *Val* with $dup_a$. We improved this merge process to be able to process three inputs—the source and the two views—by using *resC* to merge *Val*. For example, when $(N \ a \ [b]), [N \ a \ [b, c^+], N \ a \ [b^-]]$ is given as the input to *resC*, $(N \ a \ [c^+])$ results as its output. This can be explained using the manipulated positions in $c^+$ and $b^-$ in terms of *Command Val* as follows: Insert [1] $c$ and Delete [0] $b$ are transformed to produce Insert [0] $c$, so that when the operation is applied, Delete [0] $b \circ$ Insert [0] $c$ is executed. From the viewpoint of merging the updates performed on the duplicated views created by forward transformation, the redefined $\lceil Dup' \rceil^{\vee}$ acts similarly to the original $\lceil Dup \rceil^{\vee}$. When $(N \ a \ [b]), [N \ a \ [b, c^+], N \ a \ [b]]$ is given as the input, the only *Command Val* content is Insert [1] $c$, so the $N \ a \ [b, c^+]$ is produced as the result. As a BX, $\triangleleft_{Dup'}$ is defined as follows.

$$c \ \triangleleft_{Dup'} \ (Node \text{ "Dup"} \ [a_1, a_2]) =$$
$$\text{let } op_1 = \text{diff}(a_1, c) \text{ in}$$
$$\text{let } op_2 = \text{diff}(a_2, c) \text{ in}$$
$$op_1 \circ T(op_2, op_1) \ c$$

diff is a function to compute difference between the two inputs with identical data structure. The view and the source are given to diff, computing editing operations that are only present in the view, and fed to the OT as its input.

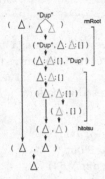

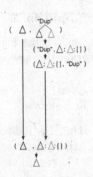

**Fig. 6.** Dataflow in ⌈Dup⌉˘          **Fig. 7.** Dataflow in ⌈Dup′⌉˘

## 8  Implementation and Property Satisfaction Evaluation

To confirm that the proposed method behaves correctly, we need to verify if both the properties of BX and OT are satisfied simultaneously. For the BX property, we conducted the following three tests for PUTGETPUT and GETPUTGET properties on a test XML data.

- OT property of *mput*: *mget*, *mput*, *mget* and *mput* are executed in this order. The test inputs of *mput* contain different new editing operations for each of the two views. We check if the conflicts among different editing operations on the two views are resolved and reflected to the source. In addition, we check if the result of the second *mput* maintains the previous updates.
- PUTGETPUT: *mput*, *mget* and *mput* are executed in order to check if no information is lost.
- GETPUTGET: *mget*, *mput* and *mget* are executed in order to check if no information is lost.

We confirmed that the proposed system satisfies, in X language and XEditor, the properties PUTGETPUT, GETPUTGET, and TP1($C_1$) at the same time. Our implementation in Haskell, as well as the test code used in the evaluation can be found at https://github.com/habutaso/x-inv-xeditor-dev.

## 9  Conclusion and Future Work

In this paper, we proposed a framework to reflect conflicting changes on views of two BX by propagating the changes to the shared source after resolving the conflict, as a step to cope with conflicts among multiple BX. The new function *mput* was introduced to BX, enabling more flexible data updates by multiple views equipped with conflict resolution that existing BX does not support. We applied OT to resolve conflicts between two views, while maintaining each update as much as possible. We implemented our method as an extension of duplication operator Dup in X/Inv, without losing the expressiveness of the original X/Inv.

We confirmed, by several testing scenarios, that well-behavedness as BX as well as $C_1$ confluence property as OT are satisfied at the same time.

Other OT with the $C_1$ property could be applied. For example, if we consider more than two views, we need the additional property $C_2$ because the final results diverge otherwise. We did not study such case because the proof of $C_2$ was omitted in the tree OT we applied [11]. We support only tree data structures. Different OT mechanism would be required for other data structures. Each user may perform more than two operations in a single revision. We would need to integrate buffering as well as handling OT of composite operations in BX. From the BX point of view, we implemented only two BX with one shared source. With two sources and two views, our method cannot be applied directly. These are left as our future work.

# References

1. The Coq proof assistant: Institut National de Recherche en Informatique et en Automatique. https://coq.inria.fr/
2. Cairns, B.: Build collaborative apps with Google Drive Realtime API. Google G Suite Developers Blog (2013)
3. Cleve, A., Kindler, E., Stevens, P., Zaytsev, V.: Multidirectional Transformations and Synchronisations (Dagstuhl Seminar 18491), vol. 8, pp. 1–48. Schloss Dagstuhl-Leibniz-Zentrum fuer Informatik (2019)
4. Diskin, Z.: Model synchronization: mappings, tiles, and categories. In: Fernandes, J.M., Lämmel, R., Visser, J., Saraiva, J. (eds.) GTTSE 2009. LNCS, vol. 6491, pp. 92–165. Springer, Heidelberg (2011). https://doi.org/10.1007/978-3-642-18023-1_3
5. Diskin, Z., Xiong, Y., Czarnecki, K.: From state- to delta-based bidirectional model transformations. In: Tratt, L., Gogolla, M. (eds.) ICMT 2010. LNCS, vol. 6142, pp. 61–76. Springer, Heidelberg (2010). https://doi.org/10.1007/978-3-642-13688-7_5
6. Ellis, C.A., Gibbs, S.J.: Concurrency control in groupware systems. In: Proceedings of the SIGMOD 1989, pp. 399–407 (1989)
7. Hu, Z., Mu, S.C., Takeichi, M.: A programmable editor for developing structured documents based on bidirectional transformations. In: Proceedings of the 2004 ACM SIGPLAN PEPM, pp. 178–189 (2004)
8. Hu, Z., Mu, S., Takeichi, M.: A programmable editor for developing structured documents based on bidirectional transformations. High. Order Symb. Comput. 21(1–2), 89–118 (2008)
9. Li, N., Tsigkanos, C., Jin, Z., Dustdar, S., Hu, Z., Ghezzi, C.: POET: privacy on the edge with bidirectional data transformations. In: Proceedings of the IEEE PerCom, pp. 1–10 (2019)
10. Mu, S.-C., Hu, Z., Takeichi, M.: An algebraic approach to bi-directional updating. In: Chin, W.-N. (ed.) APLAS 2004. LNCS, vol. 3302, pp. 2–20. Springer, Heidelberg (2004). https://doi.org/10.1007/978-3-540-30477-7_2
11. Sinchuk, S., Chuprikov, P., Solomatov, K.: Verified operational transformation for trees. In: Blanchette, J.C., Merz, S. (eds.) ITP 2016. LNCS, vol. 9807, pp. 358–373. Springer, Cham (2016). https://doi.org/10.1007/978-3-319-43144-4_22

# Entity Matching with String Transformation and Similarity-Based Features

Kazunori Sakai[1], Yuyang Dong[2(✉)], Masafumi Oyamada[2], Kunihiro Takeoka[2], and Takeshi Okadome[1]

[1] Graduate School of Science and Technology, Kwansei Gakuin University, Hyogo, Japan
`tokadome@acm.org`
[2] NEC Corporation, Tokyo, Japan
`{dongyuyang,oyamada,k_takeoka}@nec.com`

**Abstract.** Entity matching is an important task in common data cleaning and data integration problems of determining two records that refer to the same real-world entity. Many research use string similarity as features to infer entity matching but the power of the similarity may be affected by the pairs of hard-to-classify entities, which are actually different entities but have a high similarity or the same entity with low similarity. String transformation is a good solution to solve different representations between two domains of datasets, such as abbreviations, misspellings, and other expressions.

In this paper, we propose two powerful features, *similarity gain* and *dissimilarity gain*, that enables us to discriminate whether the two entities refer to the same entity after string transformation. The similarity gain is defined by the maximum amount of similarity increase among the variations in similarity before and after applying string transformations. The dissimilarity is defined by the maximum amount of similarity decrease. Moreover, the similarity gain and dissimilarity gain can also be used for selecting valuable samples in a limited labeling budget. Sufficient experiments are conducted, and our method with the proposed features improves the best accuracy in most cases.

**Keywords:** Entity matching · Entity resolution · Supervised learning · String similarity · String transformation · Feature engineering

## 1 Introduction

Entity matching is also known as record linkage, deduplication, fuzzy-join, and entity resolution. It is the task of finding records referring to the same entity across different datasets. Common entities always exist in heterogeneous datasets from different domains, such as the same product in the sales table from different EC sites. There are no unique keys to identify the same entities, and the

ⓒ Springer Nature Switzerland AG 2022
G. Fletcher et al. (Eds.): SFDI 2021, CCIS 1457, pp. 76–87, 2022.
https://doi.org/10.1007/978-3-030-93849-9_5

representation of the same entity is always different in different datasets. For example, P1 ("Digital Camera Soft Case 2") and Q1 ("DCSC2") in Table 1a are the same entity but with different representations.

**Table 1.** (a) Example records contained in the dataset $P, Q$. For example, (P1, Q1) refers to the same entity, although they are not similar. Conversely, (P2, Q2) refers to different entities, although they are similar. (b) Examples of extended records for (c) when applying string transformations that extract capital letters and strings that show product codes (strings containing numbers and capital letters).

| ID | Name |
|----|------|
| P1 | Digital Camera Soft Case 2 |
| P2 | External Hard Drive - 301302U |
| Q1 | DCSC2 |
| Q2 | External Hard Drive - 301304U |

| ID | Name | capital letters | product code |
|----|------|-----------------|--------------|
| P1 | Digital Camera Soft Case 2 | DCSC | 2 |
| P2 | External Hard Drive - 301302U | EHDU | 301302U |
| Q1 | DCSC2 | DCSC | DCSC2 |
| Q2 | External Hard Drive - 301304U | EHDU | 301304U |

(a) Records of datasets $P, Q$.     (b) Examples after string transformations.

In this paper, we focus on the similarity-based supervised learning approaches for entity matching [4,7,11]. Entity pairs are compared using various similarity functions that are applicable to their common attributes. The similarity function returns a numerical similarity value, typically normalized to $[0, 1]$, and the more similar the attributes are, the higher the value between them. However, the features of similarities may be powerless when a pair refers to the same entity but has low similarity, or a pair refers to different entities but has high similarity. This always happens in real datasets since the "left" and "right" tables in entity matching problem are from different domains, and they will have different representations, writing styles, abbreviations, and misspellings. To solve this, many research uses string transformation before matching entities [1,2,5,8,14,15,21,23]. For example, the pair of P1 and Q1 ("Digital Camera Soft Case 2," "DCSC2") in Table 1a refer to the same entity but one of the is in abbreviations. We can take the string transformation method that takes the capital characters on P1 so that makes P1 and Q1 similar. This transformation positively impacts the pair of ("Digital Camera Soft Case 2," "DCSC") which refers to the same entity. However, it also leads to a wrong answer such as ("External Hard Drive - 301302U," "External Hard Drive - 301304U"), which refers to different entities but becomes more similar after transformation. We call this kind of entity pairs as *hard-to-classify*.

To deal with the problem of the hard-to-classify pairs, two strong features, *similarity gain* and *dissimilarity gain*, are proposed that enable us to discriminate them Similarity gain and dissimilarity gain are computed by the amount of change in similarity before and after the string transformations. We add these two features as input to a supervised classification model, and we also select good training data with weighted sampling based on the similarity gain and dissimilarity gain. For testing the effectiveness, we conducted experiments on three real-world datasets. The results show that our proposed method can improve the baseline even in a little training data. The contributions of this paper are as follows:

- We propose two features, similarity gain and dissimilarity gain, that enable us to determine whether the two elements of a record pair refer to the same entity.
- We propose a train data selection method based on the similarity gain and dissimilarity gain.
- We conduct experiments on three real-world datasets, and the results suggest that our method can improve the baseline even in a little training data.

## 2   Preliminaries

In this section, we describe how to match entities with the feature vectors using string similarity functions and string transformation. Figure 1 gives a general workflow of supervised entity matching processing.

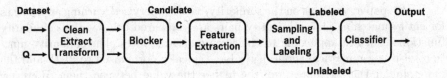

**Fig. 1.** Overview of supervised entity matching with data cleaning, blocking, feature engineering, labelling and classification.

Let $P = \{p_1, \ldots, p_{|P|}\}$ and $Q = \{q_1, \ldots, q_{|Q|}\}$ be two datasets consisting of entities. Each entity in the datasets P and Q has a common attribute $A = \{a_1, \ldots, a_{|A|}\}$ in the same domain. Let $p[a]$ be the attribute $a \in A$ of the entity $p \in P$. We assume that each attribute is represented by a string value, e.g., "Tokyo" in the *City* attribute.

Normally, there are $|P| \times |Q|$ of entities to be compared. To avoid the quadratic complexity of comparing all entity pairs, a small set of candidates are generated by the techniques of blocking [18]. For example, entities containing the same "Tokyo" in their *City* attribute can be blocked together as candidate pairs. In this work, we focus on the matching phase with given candidate pairs generated by blocking, i.e., the latter parts of feature extraction, sampling and labeling, classifier in Fig. 1.

The idea of similarity-based entity matching is to compute various similarity values with different similarity functions, and each similarity value is treated as a feature, i.e., a variable in machine learning prediction. The classifier for entity matching is trained by the feature vector composed of all similarity values. In detail, we consider a function $s$ as an indicator of how similar the two attributes (strings) are, and the value of a similarity function is normalized from 0 to 1.

$$s : A \times A \to [0, 1]. \tag{1}$$

There are many such functions, such as the Levenshtein, Jaccard, and Jaro-Winkler similarities, and we denote the set of functions as $S$. We define $S_a \subseteq S$ as a set of similarity functions applicable to attribute $a$. Thus, by applying similarity functions for an entity pair $(p, q)$, the feature vector of an entity pair is given by,

$$\mathbf{x}_{p,q} = [s\,(p[a], q[a]) \mid a \in A, s \in S_a]. \tag{2}$$

To deal with the different representation of the strings between $P$ and $Q$, we further consider the string transformation,

$$f : A \times A \to A \times A. \tag{3}$$

A string transformation function $f$ is defined to transform two attributes (strings) into another two strings, such as deletion or replacement of specific strings, capitalization, and extracting initial letters. Table 1 shows an example of applying the same transformation to both sides of the pair. However, we also support to applies different transformations to each side of the pair. Let $F$ be a set of string transformations and $F_a \subseteq F$ is applicable to attribute $a$[1]. The feature vector considering the string transformation is:

$$\mathbf{x}_{p,q} = [s\,(f\,(p[a], q[a])) \mid a \in A, s \in S_a, f \in F_a]. \tag{4}$$

## 3   Methodology

In this section, we introduce the proposed power features of similarity gain and dissimilarity gain, as well as making use of them in selecting training data.

### 3.1   Similarity Gain and Dissimilarity Gain

For a pair of entities and a specific similarity function, they are four patterns for their relationship and similarity:

---

[1] $F$ can also include function $f_0$ that does not transform anything.

- (1) Same entities with high similarity
- (2) Different entities with low similarity
- (3) Same entities with low similarity
- (4) Different entities with high similarity

The entities in (1) and (2) are easy to be classified because the similarity value is a clear signal, i.e., a power feature. However, the entities in (3) and (4) are hard to be classified and also affect the power of the similarity feature in classification.

Our goal is to design good features to deal with the hard-to-classify samples in (3) and (4). For entity pairs in (3) and (4), we observed that the string transformation is effective for entity matching when the representation is aligned to the same one after transformation. The similarity after this effective string transformation increases significantly if the two records refer to the same entity. Otherwise, the similarity decreases greatly or changes a little bit if they refer to different entities. Unfortunately, we can not know what string transformation function is effective for which similarity evaluation in every entity pair. Thus, we propose to use the amount of change in many similarities with various string transformation functions, and defined *similarity gain* and *dissimilarity gain* by aggregating all changed similarity values.

For example, assume that there is an effective string transformation that extracts capital letters for ("Digital Camera soft case 2," "DCSC2") in Table 1, as well as another string transformation that extracts the product code for ("External Hard Drive - 301302U," "External Hard Drive - 301304U"). We then obtain the transformed pairs ("DCSC," "DCSC") and ("EHDU," "EHDU") with high similarity using a string transformation that extracts capital letters (Table 1b). The latter ("EHDU," "EHDU"), however, refers to different entities. Now let us turn to the amount of change in the similarity between the similarity of the original record pairs and the similarity of the transformed pairs. We can then see that the original record pair ("Digital Camera soft case 2," "DCSC2") has a low similarity, and the amount of change in similarity after the string transformation is large. By contrast, ("External Hard Drive - 301302U," "External Hard Drive - 301304U") have a high similarity before the string transformation, and thus the change is small. By determining and selecting the effective string transformation for an entity pair based on the magnitude of the variation, we can obtain powerful features to determine whether the entity pair refers to the same one.

**Similarity Gain.** Similarity gain is a measure of how much applying the string transformations increase the string similarity of an entity pair. If two entities have low string similarity but refer to the same one, we think this case is that two different representations are derived from the same entity. Therefore, there should exist an effective transformation that links them and makes them more similar. If the prepared transformation functions contain that effective string transformation, the similarity between the transformed entities will be higher than the original one, and the similarity gain will be larger. Otherwise, the similarity gain will be less than.

Therefore, for each similarity function, we take the maximum amount of similarity gain in all transformation functions. Then, we take the average value as the final similarity gain. Formally,

$$SG(p[a], q[a]) = \frac{1}{|S_a|} \sum_{s \in S_a} \max_{f \in F_a} \left\{ s\left(f\left(p\left[a\right], q\left[a\right]\right)\right) - s\left(p\left[a\right], q\left[a\right]\right) \right\}. \quad (5)$$

**Dissimilarity Gain.** Similar to the similarity gain, the dissimilarity gain measures how much applying the string transformations decrease the string similarity of an entity pair (Fig. 2). It is defined as the average minimum value among all possible similarity functions and transformation functions. Formally,

$$DSG(p[a], q[a]) = \left| \frac{1}{|S_a|} \sum_{s \in S_a} \min_{f \in F_a} \left\{ s\left(f\left(p\left[a\right], q\left[a\right]\right)\right) - s\left(p\left[a\right], q\left[a\right]\right) \right\} \right|. \quad (6)$$

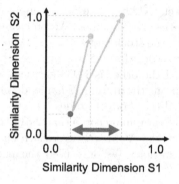 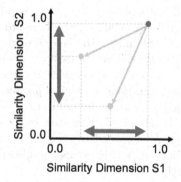

(a) The similarity gain of the similarity function S1 by the blue double-headed arrow.

(b) The dissimilarity gain of similarity functions S1 and S2 by the blue double-headed arrows.

**Fig. 2.** Similarity and dissimilarity gains (Color figure online)

## 3.2 Select Training Data with Similarity and Dissimilarity Gains

In practice, the label cost of entity matching is high so that we cannot ask to label all candidate pairs. Therefore, how to effectively select and label valuable samples is an important problem. Recall that there are four patterns of entities we introduced in Sect. 3.1, and we aim to select the samples that cover all four patterns. Therefore, we propose to select the training data by weighted sampling with four well-tuned similarity thresholds.

(1) **Same entities with high similarity.** Easy-positive samples, we select the samples by taking the entity pairs with a similarity value above $\tau_p^e$.

(2) **Different entities with low similarity.** Easy-negative samples, we select the samples by taking the entity pairs with a similarity value below $\tau_n^e$.

(3) **Same entities with low similarity.** hard-positive samples, we select the samples by taking the entity pairs with a similarity gain value above $\tau_p^h$.

(4) **Different entities with high similarity.** hard-negative samples, we select the samples by taking the entity pairs with a similarity gain value above $\tau_n^h$.

## 4   Experiments

### 4.1   Datasets

We conducted experiments on three real-world datasets to test the effectiveness of our proposed features as well as the training data selection. Details of the dataset are listed in Table 2.

– **StockName.** The dataset we collected consists of stock names and their official names in the Japanese stock market (NI225). The only comparable attribute is the name. There are 2,959 entities of companies. We processed a self-matching with their company names to the stock codes.
– **Abt-Buy.** This dataset consists of product records for two e-commerce retailers. One is 1,081 entities from abt.com and the other is 1,092 entities from buy.com. Each entity is a product that contains the attributes name, description, manufacturer, and price. We only used the product names.
– **DBLP-ACM.** This dataset consists of bibliographic records that 2,614 entities from the DBLP and 2,294 entities from the ACM websites. Each entity contains a title, venue, publication, and authors. We only used the paper titles.

We prepared 25, 18, and 15 string transformations, such as deletion and replacement of specific strings and transformation to lowercase, for the three datasets, respectively. In addition, because the StockName consisted of records in Japanese, we also prepared Japanese-only string transformations such as transforming hiragana (the Japanese syllabary characters) to Romaji.

**Table 2.** Details of datasets.

| Dataset name | Records in dataset P | Records in dataset Q | # record pairs that refer to the same entities |
|---|---|---|---|
| StockName | 2959 | 2959 | 2959 |
| Abt-Buy | 1081 | 1092 | 1097 |
| DBLP-ACM | 2614 | 2294 | 2224 |

## 4.2   Methods

We compared our proposed features as input to the classifier with the following features. We used the Random Forest as the classifier in all methods.

- **M.** Magellan [11], a string similarity-based entity matching method. The system uses a set of string similarities to acquire feature vectors and train a classifier. In our experiments, we used Jaccard, Levenstein, Jaro, Jaro-Winkler, Needleman-Wunsch, Smith-Waterman, and Monge-Elkan.
- **MT.** M+ extra features by various string transformations.
- **MTG (Ours).** MT+ our proposed similarity gain and dissimilarity gain features.

We used the blocking algorithms provided in Magellan to reduce the direct product set to a candidate set. We further randomly split the candidate set into the candidate training dataset and the test dataset. Then we selected the

**Table 3.** Comparison results in the StockName dataset. The highest average score in each training data proportion is marked bold.

| Matching methods | Sampling methods | The proportion of training data | | | | |
|---|---|---|---|---|---|---|
| | | 0.05% | 0.1% | 0.2% | 0.5% | 1% |
| M | RS | 0.101 ± 0.201 | 0.103 ± 0.196 | 0.276 ± 0.199 | 0.447 ± 0.222 | 0.522 ± 0.186 |
| | WS | 0.404 ± 0.203 | 0.528 ± 0.260 | 0.603 ± 0.233 | 0.650 ± 0.163 | 0.700 ± 0.100 |
| | WSG (Ours) | 0.437 ± 0.210 | 0.537 ± 0.103 | 0.606 ± 0.058 | 0.630 ± 0.059 | 0.674 ± 0.047 |
| MT | RS | 0.128 ± 0.256 | 0.142 ± 0.255 | 0.416 ± 0.200 | 0.590 ± 0.261 | 0.685 ± 0.190 |
| | WS | 0.509 ± 0.259 | 0.577 ± 0.251 | 0.639 ± 0.266 | 0.701 ± 0.135 | 0.749 ± 0.052 |
| | WSG (Ours) | 0.502 ± 0.228 | 0.644 ± 0.111 | 0.701 ± 0.067 | 0.714 ± 0.073 | 0.749 ± 0.064 |
| MTG (Ours) | RS | 0.122 ± 0.247 | 0.149 ± 0.270 | 0.435 ± 0.234 | 0.597 ± 0.200 | 0.680 ± 0.201 |
| | WS | 0.512 ± 0.266 | 0.564 ± 0.294 | 0.682 ± 0.274 | 0.708 ± 0.136 | 0.756 ± 0.052 |
| | WSG (Ours) | **0.597** ± 0.226 | **0.654** ± 0.048 | **0.702** ± 0.058 | **0.723** ± 0.056 | **0.757** ± 0.088 |

**Table 4.** Comparison results in the Abt-Buy dataset. The highest average score in each training data proportion is marked bold.

| Matching methods | Sampling methods | The proportion of training data | | | | |
|---|---|---|---|---|---|---|
| | | 0.05% | 0.1% | 0.2% | 0.5% | 1% |
| M | RS | 0.264 ± 0.156 | 0.260 ± 0.116 | 0.307 ± 0.089 | 0.356 ± 0.044 | 0.397 ± 0.022 |
| | WS | 0.243 ± 0.087 | 0.304 ± 0.064 | 0.347 ± 0.036 | 0.379 ± 0.027 | 0.403 ± 0.019 |
| | WSG (Ours) | 0.289 ± 0.079 | 0.310 ± 0.067 | 0.365 ± 0.031 | 0.370 ± 0.029 | 0.392 ± 0.012 |
| MT | RS | 0.494 ± 0.137 | 0.576 ± 0.127 | 0.704 ± 0.048 | 0.763 ± 0.021 | 0.782 ± 0.007 |
| | WS | 0.534 ± 0.121 | 0.647 ± 0.089 | 0.740 ± 0.019 | 0.770 ± 0.017 | 0.785 ± 0.012 |
| | WSG (Ours) | 0.554 ± 0.101 | 0.658 ± 0.042 | 0.701 ± 0.055 | 0.763 ± 0.026 | 0.795 ± 0.014 |
| MTG (Ours) | RS | 0.498 ± 0.158 | 0.585 ± 0.127 | 0.737 ± 0.030 | 0.768 ± 0.012 | 0.786 ± 0.009 |
| | WS | 0.558 ± 0.166 | 0.643 ± 0.124 | 0.724 ± 0.015 | 0.775 ± 0.012 | **0.796** ± 0.008 |
| | WSG (Ours) | **0.618** ± 0.068 | **0.701** ± 0.047 | **0.758** ± 0.040 | **0.777** ± 0.016 | 0.795 ± 0.013 |

**Table 5.** Comparison results in the DBLP-ACM dataset. The highest average score in each training data proportion is marked bold.

| Matching methods | Sampling methods | The proportion of training data | | | | |
|---|---|---|---|---|---|---|
| | | 0.05% | 0.1% | 0.2% | 0.5% | 1% |
| M | RS | 0.644 ± 0.330 | 0.877 ± 0.042 | 0.894 ± 0.025 | 0.917 ± 0.009 | 0.926 ± 0.008 |
| | WS | 0.765 ± 0.228 | 0.889 ± 0.038 | 0.908 ± 0.019 | 0.924 ± 0.019 | 0.934 ± 0.004 |
| | WSG (Ours) | 0.795 ± 0.151 | 0.852 ± 0.087 | 0.833 ± 0.139 | 0.894 ± 0.077 | 0.929 ± 0.007 |
| MT | RS | 0.691 ± 0.350 | 0.890 ± 0.042 | 0.904 ± 0.024 | 0.925 ± 0.010 | 0.937 ± 0.007 |
| | WS | 0.855 ± 0.108 | **0.915 ± 0.017** | **0.924 ± 0.012** | 0.933 ± 0.009 | 0.941 ± 0.005 |
| | WSG (Ours) | 0.851 ± 0.119 | 0.828 ± 0.122 | 0.877 ± 0.101 | 0.909 ± 0.079 | 0.919 ± 0.079 |
| MTG (Ours) | RS | 0.714 ± 0.351 | 0.886 ± 0.048 | 0.908 ± 0.025 | 0.929 ± 0.011 | 0.936 ± 0.006 |
| | WS | 0.851 ± 0.140 | 0.914 ± 0.023 | 0.923 ± 0.012 | **0.934 ± 0.006** | **0.942 ± 0.005** |
| | WSG (Ours) | **0.880 ± 0.097** | 0.891 ± 0.132 | 0.904 ± 0.086 | 0.922 ± 0.041 | 0.941 ± 0.009 |

actual training data to be labeled from the candidate training dataset with the following methods.

– **RS.** Random sampling
– **WS.** Weighted sampling on cases (1) and (2) in Sect. 3.2 by the Levenstein similarity.
– **WSG (Ours).** Weighted sampling on all cases (1), (2), (3) and (4) in Sect. 3.2 by the Levenstein similarity.

### 4.3   Evaluation Metrics

To evaluate the performance of different methods, we compared their ability to reduce the number of misclassifications. Misclassification includes false positive (record pairs that refer to the same entities but classified as not) and false negative (record pairs that refer to different entities but classified as not). In the experiment, we used the F1 measure to take into account both precision and recall. It has a maximum value of 1 and a worst-case value of 0, which is calculated by,

$$F_1 = \frac{2 \times \text{precision} \times \text{recall}}{\text{precision} + \text{recall}}. \tag{7}$$

### 4.4   Results

We vary the proportion of the training data from 0.05% to 1.0% and report the average of 50 training results. We tuned the parameters $\tau_p^e$, $\tau_n^e$, $\tau_p^h$, and $\tau_n^h$ in our training data selecting method as 0.4, 0.35, 0.4 and 0.4, respectively.

Table 3 shows the comparison results on the StockName dataset. Our method with two features of similarity gain and dissimilarity gain and training selection improve the most accuracy. Let us here focus on the methods of training data selection. While RS is very low, WS and WSG (Ours) significantly improved the accuracy of the classification using any features (M, MT, MTS (Ours)) as input.

Furthermore, using our proposed features also makes the classifier have better accuracy than WS (bottom row of Table 3).

Table 4 shows the comparison results on the Abt-Buy dataset. The results show the same trend as the results of the StockName dataset. Using our proposed features for both classifier features and selecting training data improved the classification accuracy when the proportion of the training data is small, i.e., <0.5%.

Table 5 shows the comparison results on the DBLP-ACM dataset. Our method had the best performance in a small proportion of 0.05%, but failed to be the best in a large proportion. This is because the DBLP-ACM dataset is as ambiguous as the others two datasets, the titles of the papers do not contain abbreviations, misspellings, and synonyms. Therefore, the proportion of entity pairs that refer to different entities with high similarity, or that of refer to the same entity with low similarity, is small. The experimental results suggest that using string transformations on the titles of papers is not an effective way of entity matching.

# 5   Related Work

**Similarity-Based Supervised Learning for Entity Matching.** Many recent record linkage methods focus on supervised learning approaches that require labeled training data. Superior supervised learning methods, such as tree-based models and support vector machines, have been applied in the field of record linkage [4,21]. Supervised learning methods generally produce better results than unsupervised ones, but it is challenging to prepare training data in most real-world situations. Christen [6] proposed a training data selection method using similarities. Arasu et al. [3] and Sarawagi et al. [20] proposed reducing candidate record pairs by using string similarities and selecting informative record pairs by active learning using committees. Refinement of the similarity measure, which represents whether a record pair refers to the same entity, is also being intensively studied. As mentioned above, since similarity measures are also used for training data selection, discovering good them is a significant challenge for record linkage. Fellegi et al. [10] represented a record pair as a binary vector consisting of whether the rules are satisfied or not Cohen et al. [7] proposed an adaptive framework that combines multiple similarity measures. Methods for learning similarity measures are also proposed. Bilenko et al. [4] used SVM to create a domain-adapted similarity measure. The studies of [1,2,5,14,15,21] mentioned that string transformations resolve variations of representation such as abbreviations and misspelling.

**Deep Entity Matching.** Many studies use deep learning for entity matching. DeepER [9] and DeepMatcher [16] uses a RNN model with word embeddings. Auto-EM [22] and Ditto [13] leverage a transfer learning with pre-trained language models for efficient entity matching. We do not compare the deep learning approaches since they are orthogonal to our work, and we focus on improving the accuracy of the lightweight models with similarity-based features.

**Entity Matching Systems.** Recently, there are also systems for automatic entity matching with less human involvement cost. Magellan [11] is the state-of-the-art similarity feature-based entity matching system. AutoML-EM [17,19] studies how to automatically build a machine learning model for entity matching using the techniques of AutoML. Auto-FuzzyJoin [12] is an unsupervised approach that automatically tunes the parameter for preprocessing methods, tokenizations, and similarity functions.

## 6    Conclusion

In this study, we proposed and evaluated the features, the similarity gain and the dissimilarity gain, for robust entity matching with similarity-based approaches. We also make use of the similarity gain and the dissimilarity gain in selecting valuable samples for training data with a lower labeling cost. In the experiments over three real-world datasets, our proposed features improved the accuracy even in the case that the proportion of the training data was small. In the future, we plan to generate more various and powerful features for entity matching based on the similarity gain and the dissimilarity gain.

## References

1. Arasu, A., Chaudhuri, S., Kaushik, R.: Transformation-based framework for record matching. In: 2008 IEEE 24th International Conference on Data Engineering, pp. 40–49. IEEE (2008)
2. Arasu, A., Chaudhuri, S., Kaushik, R.: Learning string transformations from examples. Proc. VLDB Endow. **2**(1), 514–525 (2009)
3. Arasu, A., Götz, M., Kaushik, R.: On active learning of record matching packages. In: Proceedings of the 2010 ACM SIGMOD International Conference on Management of data, pp. 783–794 (2010)
4. Bilenko, M., Mooney, R.J.: Adaptive duplicate detection using learnable string similarity measures. In: Proceedings of the Ninth ACM SIGKDD International Conference on Knowledge Discovery and Data Mining, pp. 39–48 (2003)
5. Çakal, Ö.Ö., Mahdavi, M., Abedjan, Z.: CLRL: feature engineering for cross-language record linkage. In: EDBT, pp. 678–681 (2019)
6. Christen, P.: Automatic training example selection for scalable unsupervised record linkage. In: Washio, T., Suzuki, E., Ting, K.M., Inokuchi, A. (eds.) PAKDD 2008. LNCS (LNAI), vol. 5012, pp. 511–518. Springer, Heidelberg (2008). https://doi.org/10.1007/978-3-540-68125-0_45
7. Cohen, W.W., Richman, J.: Learning to match and cluster large high-dimensional data sets for data integration. In: Proceedings of the Eighth ACM SIGKDD International Conference on Knowledge Discovery and Data Mining, pp. 475–480 (2002)
8. Deng, D., et al.: Unsupervised string transformation learning for entity consolidation. In: 35th IEEE International Conference on Data Engineering, ICDE 2019, Macao, China, 8–11 April 2019, pp. 196–207. IEEE (2019)
9. Ebraheem, M., Thirumuruganathan, S., Joty, S.R., Ouzzani, M., Tang, N.: Distributed representations of tuples for entity resolution. Proc. VLDB Endow. **11**(11), 1454–1467 (2018)

10. Fellegi, I.P., Sunter, A.B.: A theory for record linkage. J. Am. Stat. Assoc. **64**(328), 1183–1210 (1969)
11. Konda, P., Das, S., Suganthan G.C.P., Doan, A., Ardalan, A., et al.: Magellan: Toward building entity matching management systems
12. Li, P., Cheng, X., Chu, X., He, Y., Chaudhuri, S.: Auto-fuzzyjoin: auto-program fuzzy similarity joins without labeled examples (2021)
13. Li, Y., Li, J., Suhara, Y., Doan, A., Tan, W.: Deep entity matching with pre-trained language models. Proc. VLDB Endow. **14**(1), 50–60 (2020)
14. Michelson, M., Knoblock, C.A.: Mining the heterogeneous transformations between data sources to aid record linkage. In: ICAI (2009)
15. Minton, S.N., Nanjo, C., Knoblock, C.A., Michalowski, M., Michelson, M.: A heterogeneous field matching method for record linkage. In: Fifth IEEE International Conference on Data Mining (ICDM'05). 8p. IEEE (2005)
16. Mudgal, S., et al.: Deep learning for entity matching: a design space exploration. In: Das, G., Jermaine, C.M., Bernstein, P.A. (eds.) Proceedings of the 2018 International Conference on Management of Data, SIGMOD Conference 2018, Houston, TX, USA, 10–15 June 2018. pp. 19–34. ACM (2018)
17. Paganelli, M., Buono, F.D., Pevarello, M., Guerra, F., Vincini, M.: Automated machine learning for entity matching tasks. In: Velegrakis, Y., Zeinalipour-Yazti, D., Chrysanthis, P.K., Guerra, F. (eds.) Proceedings of the 24th International Conference on Extending Database Technology, EDBT 2021, Nicosia, Cyprus, 23–26 March 2021, pp. 325–330. OpenProceedings.org (2021)
18. Papadakis, G., Skoutas, D., Thanos, E., Palpanas, T.: A survey of blocking and filtering techniques for entity resolution. CoRR abs/1905.06167 (2019). http://arxiv.org/abs/1905.06167
19. Wang, P., Zheng, W., Wang, J., Pei, J.: Automating entity matching model development. In: 37th IEEE International Conference on Data Engineering, ICDE 2021. IEEE (2021)
20. Sarawagi, S., Bhamidipaty, A.: Interactive deduplication using active learning. In: Proceedings of the Eighth ACM SIGKDD International Conference on Knowledge Discovery and Data Mining, pp. 269–278 (2002)
21. Tejada, S., Knoblock, C.A., Minton, S.: Learning domain-independent string transformation weights for high accuracy object identification. In: Proceedings of the Eighth ACM SIGKDD International Conference on Knowledge Discovery and Data Mining, pp. 350–359 (2002)
22. Zhao, C., He, Y.: Auto-EM: end-to-end fuzzy entity-matching using pre-trained deep models and transfer learning. In: Liu, L., et al. (eds.) The World Wide Web Conference, WWW 2019, San Francisco, CA, USA, 13–17 May 2019, pp. 2413–2424. ACM (2019)
23. Zhu, E., He, Y., Chaudhuri, S.: Auto-join: Joining tables by leveraging transformations. Proc. VLDB Endow. **10**(10), 1034–1045 (2017). http://www.vldb.org/pvldb/vol10/p1034-he.pdf

# Towards Automatic Synthesis of View Update Programs on Relations

Bach Nguyen Trong[1,2(✉)] and Zhenjiang Hu[2,3]

[1] The Graduate University for Advanced Studies, SOKENDAI, Kanagawa, Japan
[2] National Institute of Informatics, Tokyo, Japan
bach@nii.ac.jp
[3] Peking University, Beijing, China
huzj@pku.edu.cn

**Abstract.** Automatic synthesis of bidirectional programs on relations has not been well studied yet. As an attempt to solve the problem, we propose an approach to synthesizing view update strategies on relations written in Datalog from examples and data schemes. Our approach has been implemented and used to successfully synthesize various view update tasks on relations.

**Keywords:** View update · Program synthesis

## 1 Introduction

Bidirectional programming (programming with bidirectional languages) is difficult due to strict constraints imposed on those languages [3]. To overcome this difficulty, many automatic synthesis approaches have been proposed [2–4]. However, they mainly focus on obtaining bidirectional regular expressions for textual data processing and it is unknown yet how to synthesize bidirectional programs on relations, which is practically important to solve the view updating problem in the database community.

In this paper, we report our first attempt of automatic synthesis of bidirectional programs on relations. More specifically, since bidirectional programs on relations can be programmed as view update programs (view update strategies) on relations using a subset of Datalog [6], we shall show how we can synthesize view update programs from examples and data schemes.

Our work is inspired by the recent synthesis system PROSYNTH [5] that can synthesize Datalog programs from examples in the domain of relations. The key idea is to reduce the synthesis of Datalog programs (a set of rules) to the rule selection from a set of candidate rules based on given examples. The challenge to use this method is to provide a set of candidate rules for the problems one

---

We thank the anonymous reviewers for their valuable feedback. This work has been partially supported by JSPS KAKENHI Grant Number JP17H06099 and ROIS NII Open Collaborative Research 2018.

G. Fletcher et al. (Eds.): SFDI 2021, CCIS 1457, pp. 88–95, 2022.
https://doi.org/10.1007/978-3-030-93849-9_6

**Table 1.** Original source database and view

| eid | ename | lid | eteam | erole |
|-----|-------|-----|-------|-------|
| 1 | Bob | 3 | A | Mgr |
| 2 | Carol | 1 | B | Dev |
| 3 | Ted | 2 | B | Dev |
| 4 | Alice | 1 | B | Mgr |

(a) Emp

| lid | laddr | lcode |
|-----|-------|-------|
| 1 | Tokyo | 81 |
| 2 | Hanoi | 84 |
| 3 | Beijing | 86 |
| 4 | Seoul | 82 |

(b) Loc

| eid | ename | lid | laddr | eteam |
|-----|-------|-----|-------|-------|
| 2 | Carol | 1 | Tokyo | B |
| 3 | Ted | 2 | Hanoi | B |
| 4 | Alice | 1 | Tokyo | B |

(c) Teamview

**Table 2.** Updated source database and updated view

| eid | ename | lid | eteam | erole |
|-----|-------|-----|-------|-------|
| ~~1~~ | ~~Bob~~ | ~~3~~ | ~~A~~ | ~~Mgr~~ |
| 1 | Mark | 1 | B | Ukn |
| ~~2~~ | ~~Carol~~ | ~~1~~ | ~~B~~ | ~~Dev~~ |
| 3 | Ted | 2 | B | Dev |
| 4 | ~~Alice~~ | ~~1~~ | ~~B~~ | ~~Mgr~~ |
| 4 | Alice | 1 | A | Mgr |
| 5 | Jack | 5 | B | Dev |

(a) Emp_upd

| lid | laddr | lcode |
|-----|-------|-------|
| 1 | Tokyo | 81 |
| 2 | Hanoi | 84 |
| 3 | Beijing | 86 |
| 4 | Seoul | 82 |
| 5 | London | 44 |

(b) Loc_upd

| eid | ename | lid | laddr | eteam |
|-----|-------|-----|-------|-------|
| 1 | Mark | 1 | Tokyo | B |
| 2 | ~~Carol~~ | ~~1~~ | ~~Tokyo~~ | B |
| 3 | Ted | 2 | Hanoi | B |
| 4 | ~~Alice~~ | ~~1~~ | ~~Tokyo~~ | B |
| 5 | Jack | 5 | London | B |

(c) Teamview_upd

wish to solve. For our purpose, we need to provide a set of candidate rules for representing all view update programs on relations.

Fortunately, about 30 years ago, when considering the problem of updating source databases though views defined by SPJNF[1] queries of relations, Keller [1] showed that it is possible to enumerate all view update strategies for the view definition queries satisfying some criteria. This hints us to design a set of candidate rules from this (informal) enumeration of view update strategies.

The main contributions of our work include (1) a new approach to synthesizing view update programs on relations from examples and schemes; (2) design of template rules for view update strategies based on Keller's enumeration of view update strategies; (3) an algorithm for generating from the template rules a set of candidate rules that can be used as the input to PROSYNTH; and (4) an implementation of our approach and a preliminary evaluation on a variety of synthesis tasks for the view update problem.

## 2 Overview

We shall present an overview of our approach using a motivating example.

**Source and View.** Let us consider a source database that contains two BCNF[2] relations Emp and Loc with their schemas as listed below:

---

[1] Selection Projection Join Normal Form.
[2] Boyce-Codd Normal Form.

---

**Algorithm 1:** SYNTHVUP($S$, $E$) - Synthesizing view update programs.

---

**Input**: $S = \{* \ sources, 1 \ view\}$, $E = (es, ev, ev_{upd}, es_{upd})$

**Output**: A satisfying program $P$ or NONE if no solution exists.

1 $P_c \leftarrow$ GENRULE($S, E$);
2 $P \leftarrow$ PROSYNTH($E, P_c$);
3 **return** $P$;

---

```
Emp (eid:EID, ename:ENAME, lid:LID, eteam:ETEAM, erole:EROLE)
Loc (lid:LID, laddr:LADDR, lcode:LCODE)
```

The relation `Emp` stores employee information records including employee's ID number, name, work location ID number, workplace team and role. `Emp` has a primary key `eid` and a foreign key `lid` that references to the primary key `lid` of the relation `Loc`. `Loc` stores a list of work locations, in which each location has its own location ID number, address and country code.

Over the source database, one may define a view `Teamview`

```
Teamview (eid:EID, ename:ENAME, lid:LID, laddr:LADDR, eteam:ETEAM)
```

which is used to extract information of Team B by the following view definition:

```
Teamview(eid, ename, lid, laddr, eteam) :-
    Emp(eid, ename, lid, eteam, _), Loc(lid, laddr, _), eteam = "B".
```

The view `Teamview` should keep all primary keys of relations `Emp` and `Loc`, as well as the names of attributes. Example instances of the relations `Emp` and `Loc` and the view `Teamview` are shown in Table 1a, Table 1b and Table 1c respectively.

**View Update Strategy.** A view update strategy describes a way to reflect changes on the view to those on the source. Suppose that the Teamview in Table 1c is changed to Table 2c that is represented by `Teamview_upd`. The black, red and blue tuples are original, inserted and deleted records respectively. We may describe our view update intention by showing an example of how the source is updated as described by Table 2a and Table 2b that are respectively represented by `Emp_upd` and `Loc_upd`. How can we automatically synthesize a view update strategy in Datalog that is consistent with the intention described by the example in Tables 1 and 2?

In fact, our system can automatically synthesize the following view update program from the above example:

```
Teamview_ins(v0,v1,v2,v3,v4) :-
    !Teamview(v0,v1,v2,v3,v4), Teamview_upd(v0,v1,v2,v3,v4).
Loc_upd(v0,v1,v2) :- Loc(v0,v1,v2).
Loc_upd(v0,v1,v2) :- !Loc(v0,_,_),
    Teamview_ins(_,_,v0,v1,_), v2="44", v0="5".
```

Readers can find a full synthesized program for the motivating example in Appendix A.

# 3   Synthesizing View Update Programs

Algorithm 1 summarizes our synthesis approach. SYNTHVUP takes a set $S$ containing one or more BCNF source schemas and one BCNF view schema whose primary key is a set of primary keys of the source schemas, and an example $E$ containing instances of tables of the sources, the view, the updated view and the updated sources, as input. First, SYNTHVUP calls GENRULE to generate a program $P_c$ that has a list of candidate rules and is useful to input PROSYNTH for doing the synthesis task. If PROSYNTH returns NONE then there is no subset of candidate rules that could be a valid solution. Otherwise, the output program $P$ of PROSYNTH($E$, $P_c$) will be a view update program that is consistent with ($S$, $E$). Since PROSYNTH is clearly presented in [5], we shall focus on the procedure GENRULE.

**Template Rules.** Following the idea of the complete enumeration of view update strategies in [1], we can construct a set of template rules from the schemes of the source and the view.

First, from the schemas of the source and the view, we can easily derive schemas of relations source_upd and view_upd.

Next, we introduce the following two relations view_del and view_ins which have the same schema with view to compute deleted and inserted view tuples repectively (vkeys and vnkeys (skeys and snkeys) are corresponding to sets of primary key and non-key attributes of the view (source)):

```
view_del(vkeys, vnkeys) :- view(vkeys, vnkeys), !view_upd(vkeys, vnkeys).
view_ins(vkeys, vnkeys) :- !view(vkeys, vnkeys), view_upd(vkeys, vnkeys).
```

Finally, we provide core template rules for writing source_upd from source, view_del and view_ins as follows:

```
source_upd(skeys, snkeys) :-
[1]. source(skeys, snkeys).
[2]. source(skeys, snkeys), !view_del(skeys, _).
[3]. source(skeys, snkeys), !view_ins(skeys, _).
[4]. source(skeys, snkeys), !view_del(skeys, _), !view_ins(skeys, _).
[5]. source(skeys, _), view_ins(skeys, subset_snkeys, _),
     [attrNK = "valueNK" for each attrNK in snkeys - subset_snkeys],
     [attrK = "valueK" for each attrK in skeys].
[6]. !source(skeys, _), view_ins(skeys, subset_snkeys, _),
     [attrNK = "valueNK" for each attrNK in snkeys - subset_snkeys],
     [attrK = "valueK" for each attrK in skeys].
[7]. source(skeys, snkeys - selecting_attr, _), view_del(skeys, _),
     [attrS = "ivalueS" for each attrS in selecting_attr],
     [attrK = "valueK" for each attrK in skeys].
```

Intuitively, Template [1] maintains the source table if there is no reflected update. Templates [2], [3] and [4] keep all tuples in the source that has no key-conflict with any updated view tuples. Templates [5], [6] and [7] handle the translations where some missing attributes need to be set.

**GenRule.** Algorithm 2 describes the procedure GENRULE($S, E$) for generating program $P_c$ that stores candidate rules.

---

**Algorithm 2:** GENRULE($S, E$) - Generating candidate rules

**Input**: $S = \{* \ sources, 1 \ view\}$, $E = (es, ev, es_{upd}, ev_{upd})$
**Output**: Program containing candidate rules $P_c$

1  $P_c \leftarrow$ empty file;
2  $s, v \leftarrow S[sources], S[view]$;
3  **for** $t \in r.type \wedge r \in s$ **do** write($P_c, t$);
4  **for** $r \in s \vee v$ **do**
5  $\quad$ $uex[r]_{del} \ (uex[r]_{ins}) \leftarrow$ load deleted (inserted) tuples against $r$ from $E$
6  $\quad$ write($P_c$, declarations of $r$ and $r\_upd$);
7  $\quad$ **if** $r == v$ **then** write($P_c$, declarations and rules of $v\_ins$ and $v\_del$)
8  **for** $r \in s$ **do**
9  $\quad$ $cr \leftarrow$ rule generated by templates [1], [2], [3], [4];
10 $\quad$ write($P_c, cr$);
11 $\quad$ $cr \leftarrow$ basis of rule generated by templates [5], [6];
12 $\quad$ $dattr \leftarrow$ set of difference attributes of $r$ against $v$;
13 $\quad$ **if** $dattr$ *is empty* **then**
14 $\quad\quad$ write($P_c, cr$);
15 $\quad$ **else**
16 $\quad\quad$ update $cr$ using $uex[r]_{ins}$ and $r.pkey$;
17 $\quad\quad$ write($P_c, cr$);
18 $\quad$ $cr \leftarrow$ basis of rule generated by templates [7];
19 $\quad$ $kval \leftarrow$ set of $r.pkey$ values of $e \in uex[r]_{del}$;
20 $\quad$ **for** $k \in kval$ **do**
21 $\quad\quad$ $(e_{del}, e_{ins}) \leftarrow$ tuples in $uex[r]_{del}, uex[r]_{ins}$ that have same key value $k$;
22 $\quad\quad$ $sels \leftarrow$ set of difference attributes of $e_{ins}$ against $e_{del}$;
23 $\quad\quad$ update $cr$ using $sels$ and $r.pkey$;
24 $\quad\quad$ write($P_c, cr$);
25 **return** $P_c$

---

First, GENRULE initializes $P_c$ as an empty file. Then, it in turn writes to $P_c$ information about types of relations in schema set $S$, declarations of source, source_upd, view and view_upd, declarations and rules of view_del and view_ins. Updated tuples for each relation are also loaded in order to be able to support the generation of candidate rules $cr$.

Next, for each relation in source databases, GENRULE generates $cr$ based on core templates and writes it to $P_c$. The value of missing attributes after updating could be got by extracting information from loaded tuples and the primary key of each relation.

Finally, GENRULE returns $P_c$ if there is no more candidate rule that could be generated from templates.

## 4    Evaluation

We have fully implemented SYNTHVUP in lines of Python code with the integration of PROSYNTH. To evaluate the success and effectiveness of synthesizing view update tasks, we constructed our benchmarks by choosing BCNF schemas, SPJ queries used for view definition and view update and their instances from [6] - an in-depth research that has already collected multiple cases from various sources and by handcrafting specifications to test the features of the algorithm.

**Table 3.** Evaluation results

| ID | Benchmark | Type | # DelTpl | # InsTpl | # CandRl | # SolRl | SynthTime (s) |
|----|-----------|------|----------|----------|----------|---------|---------------|
| 1 | car_master | P | 2 | 1 | 10 | 2 | 0.0278 |
| 2 | goodstudents | SP | 3 | 0 | 12 | 2 | 0.0444 |
| 3 | goodstudents2 | SP | 3 | 0 | 14 | 4 | 0.0445 |
| 4 | luxuryitems | S | 1 | 0 | 7 | 2 | 0.0361 |
| 5 | usa_city | SP | 0 | 1 | 7 | 2 | 0.0364 |
| 6 | paramountmovies | SP | 0 | 1 | 5 | 2 | 0.0195 |
| 7 | officeinnfo | P | 1 | 1 | 7 | 2 | 0.0279 |
| 8 | bstudents | SPJ | 0 | 2 | 10 | 5 | 0.0197 |
| 9 | location | SPJ | 1 | 2 | 21 | 7 | 0.0647 |
| 10 | department | J | 2 | 1 | 13 | 4 | 0.0552 |
| 11 | all_cars | J | 2 | 1 | 11 | 4 | 0.0442 |
| 12 | Teamview | SPJ | 2 | 2 | 17 | 6 | 0.0399 |

Table 3 shows evaluation results[3]. All constructed benchmarks are automatically synthesized by SYNTHVUP. The number of rules in a solution (#SolRl) is relatively smaller than the number of candidate rules (#CandRl) which depends on the number of updated tuples (#DelTpl, #InsTpl) in the view. View update programs for our benchmarks are found in a short time partly because of a not large #CandRl. We may need to experiment with larger scenarios in the future.

## 5    Conclusion

In this paper, we briefly explain a new approach to synthesizing a class of view update programs from examples and schemas using a Datalog synthesis system PROSYNTH as a plug-in of our synthesis algorithm. Based on a set of core template rules we develop from an old study in the view update problem, our algorithm first generates a program containing candidate rules that can then be used as an input parameter, besides examples, for PROSYNTH. We have implemented our proposed algorithm and evaluated it on a range of practical tasks

---

[3] All experiments were performed on a 2.6 GHz Intel Core i7 processor with 16 GB of 2400 MHz DDR4 running macOS 11.3.1.

where view update programs were synthesized automatically from examples and schemas.

There are some future directions towards the synthesis of view update programs or bidirectional programs on relations: (1) enriching template rules by relaxing key constraints and adding other query operations (union, intersection, set difference); (2) handling situations of multiple examples for module testing and multiple solutions where a ranking mechanism is necessary; and (3) introducing an interactive mode that allows users to get more involved in the synthesis process.

## A     A Synthesized Program of the Motivating Example

```
// subtype information
 .type ENAME <: symbol
 .type LADDR <: symbol
 .type LCODE <: symbol
 .type LID <: symbol
 .type EROLE <: symbol
 .type ETEAM <: symbol
 .type EID <: symbol

// declarations for sources
 .decl Emp(v0:EID, v1:ENAME, v2:LID, v3:ETEAM, v4:EROLE)
 .decl Emp_upd(v0:EID, v1:ENAME, v2:LID, v3:ETEAM, v4:EROLE)
 .decl Loc(v0:LID, v1:LADDR, v2:LCODE)
 .decl Loc_upd(v0:LID, v1:LADDR, v2:LCODE)
 .input Emp, Loc
 .output Emp_upd, Loc_upd

// declarations for view
 .decl Teamview(v0:EID, v1:ENAME, v2:LID, v3:LADDR, v4:ETEAM)
 .decl Teamview_upd(v0:EID, v1:ENAME, v2:LID, v3:LADDR, v4:ETEAM)
 .input Teamview, Teamview_upd

// declartions and necessary rules for view_ins and view_del
 .decl Teamview_ins(v0:EID, v1:ENAME, v2:LID, v3:LADDR, v4:ETEAM)
 .decl Teamview_del(v0:EID, v1:ENAME, v2:LID, v3:LADDR, v4:ETEAM)
 .output Teamview_ins, Teamview_del

0a. Teamview_del(v0,v1,v2,v3,v4) :-
    Teamview(v0,v1,v2,v3,v4), !Teamview_upd(v0,v1,v2,v3,v4).
0b. Teamview_ins(v0,v1,v2,v3,v4) :-
    !Teamview(v0,v1,v2,v3,v4), Teamview_upd(v0,v1,v2,v3,v4).

// solution rules synthesized from candidate rules
 1. Emp_upd(v0,v1,v2,v3,v4) :- Emp(v0,v1,v2,v3,v4),
    !Teamview_del(v0,_,_,_,_), !Teamview_ins(v0,_,_,_,_).
 2. Emp_upd(v0,v1,v2,v3,v4) :- Emp(v0,v1,v2,_,v4),
```

```
  Teamview_del(v0,_,_,_,_), v3="A", v0="4".
3. Emp_upd(v0,v1,v2,v3,v4) :- Emp(v0,_,_,_,_),
   Teamview_ins(v0,v1,v2,_,v3), v4="Ukn", v0="1".
4. Emp_upd(v0,v1,v2,v3,v4) :- !Emp(v0,_,_,_,_),
   Teamview_ins(v0,v1,v2,_,v3), v4="Dev", v0="5".
5. Loc_upd(v0,v1,v2) :- Loc(v0,v1,v2).
6. Loc_upd(v0,v1,v2) :- !Loc(v0,_,_),
   Teamview_ins(_,_,v0,v1,_), v2="44", v0="5".
```

# References

1. Keller, A.M.: Algorithms for translating view updates to database updates for views involving selections, projections, and joins. In: Proceedings of the Fourth ACM SIGACT-SIGMOD Symposium on Principles of Database Systems, PODS 1985, pp. 154–163. Association for Computing Machinery, New York (1985)
2. Maina, S., Miltner, A., Fisher, K., Pierce, B.C., Walker, D., Zdancewic, S.: Synthesizing quotient lenses. In: Proceedings of the ACM on Programming Languages, 2, pp. 1–29, July 2018
3. Miltner, A., Fisher, K., Pierce, B.C., Walker, D., Zdancewic, S.: Synthesizing bijective lenses. In: Proceedings of the ACM on Programming Languages, vol. 2, December 2017
4. Miltner, A., Maina, S., Fisher, K., Pierce, B.C., Walker, D., Zdancewic, S.: Synthesizing symmetric lenses. In: Proceedings of the ACM on Programming Languages, vol. 3, July 2019
5. Raghothaman, M., Mendelson, J., Zhao, D., Naik, M., Scholz, B.: Provenance-guided synthesis of datalog programs. In: Proceedings of the ACM on Programming Languages, vol. 4, December 2019
6. Tran, V.D., Kato, H., Hu, Z.: Programmable view update strategies on relations. Proc. VLDB Endow. 13(5), 726–739 (2020)

# Adaptive SQL Query Optimization in Distributed Stream Processing: A Preliminary Study

Darya Sharkova[1], Alexander Chernokoz[2], Artem Trofimov[3(✉)],
Nikita Sokolov[3], Ekaterina Gorshkova[4], Igor Kuralenok[3], and Boris Novikov[1]

[1] HSE University, Saint Petersburg, Russia
dmsharkova@edu.hse.ru, borisnov@acm.org
[2] ITMO University, Saint Petersburg, Russia
288467@niuitmo.ru
[3] Yandex, Saint Petersburg, Russia
{tomato,faucct,solar}@yandex-team.ru
[4] JResearch Software, Prague, Czech Republic
cathy@jresearch.org

**Abstract.** Distributed stream processing is widely adopted for real-time data analysis and management. SQL is becoming a common language for robust streaming analysis due to the introduction of time-varying relations and event time semantics. However, query optimization in state-of-the-art stream processing engines (SPEs) remains limited: runtime adjustments to execution plans are not applied. This fact restricts the optimization capabilities because SPEs lack the statistical data properties before query execution begins. Moreover, streaming queries are often long-lived, and these properties can change over time.

Adaptive optimization, used in databases for queries with insufficient or unknown data statistics, can fit the streaming scenario. In this work, we explore the main challenges that SPEs face during the adjustment of adaptive optimization, such as predicting statistical properties of streams and execution graph migration. We demonstrate potential performance gains of our approach within an extension of the *NEXMark* streaming benchmark and outline our further work.

## 1 Introduction

Modern data analytics commonly requires near real-time online processing of continuously changing data arriving from unbounded streams. A standard way of defining a stream processing pipeline is an execution graph. Each node represents an operation performed over stream elements. Although a declarative approach to defining computations dominates analytical processing (e.g., OLAP), it has not gained wide use in streaming systems. The SQL language, typically used for database queries, is a dominating means for declarative specification of

G. Fletcher et al. (Eds.): SFDI 2021, CCIS 1457, pp. 96–109, 2022.
https://doi.org/10.1007/978-3-030-93849-9_7

data processing. The advantages of SQL include its popularity and ease of adoption and its support of windowed aggregations and joins, among other highly expressive features.

Implementations of different variations of SQL language for stream processing have been an area of active development for the last two decades. However, there have not been many productive attempts at proposing a standard for robust streaming SQL. One such attempt [6] is pretty recent (2019); its predecessors, such as CQL [1], have not found much popularity. Modern stream processing engines (SPEs) typically implement only a subset of SQL features. We expect that with the recent efforts in providing a standard for streaming SQL, the declarative approach to stream processing gains popularity.

Optimization is one of the stages of executing a declarative query. First, a query is *parsed* into an abstract syntax tree, each node of which represents a relational algebra operator. Second, the query is simplified during the *rewriting* phase into a logical plan (or graph (we will be using these two terms interchangeably in the paper). Third, the optimizer builds a physical plan equivalent to the logical plan. Finally, an executor interprets the query and delivers the result to the user [25]. The correspondence between queries and plans is not one-to-one. Several execution plans can exist for any query. The purpose of query rewriting is to reduce the space of execution plans and to standardize and simplify the query for further processing [26]. As for optimization, the query planner transforms each logical operator into its physical implementation. For example, the optimizer can implement a join operation using the hash join algorithm or the merge join algorithm [23]. A join operation in a distributed system might require re-sharding. Database query optimization is a well-studied topic [2,11,13].

Modern state-of-the-art optimizers are cost-based: an optimizer builds a physical plan that minimizes the cost function, which encapsulates the complexity of processing the query. The cost function is typically a linear combination of expected I/O and CPU costs, with CPU costs of each operation estimated based on relation cardinality and operator selectivity.

The cost estimation produced by a cost function depends on statistical properties of data (such as cardinalities and frequencies of attribute values). Therefore, cost-based optimization requires knowledge of statistical information [23]. Typically database systems collect statistics periodically and use them during the optimization of incoming queries until the next scheduled update of statistics. This practice works because data are relatively stable, and statistics are not changing rapidly. In contrast, queries are optimized independently from each other.

This assumption is not valid for streams, thus obtaining such statistics in streaming systems presents specific difficulties. However, queries are executed repeatedly for a series of stream windows. In other words, queries are relatively stable while data are volatile.

Efforts to optimize streaming query execution in a distributed environment focus on finding a suitable mapping from a logical graph to a physical graph executed on a machine [9,10,12,20,28]. These optimizations are local because the system derives all candidate physical plans from the same logical plan.

In this paper, we introduce new techniques for the global optimization of declarative queries over data streams. In order to obtain the statistics needed for optimization, we use predictions based on data collected during previous executions of the same query. Further, our techniques are adaptive because they are robust to a drift of stream properties.

The contributions of this paper are the following:

- We present a detailed analysis of the problem of SQL queries optimization in distributed stream processing and discuss challenges that arise within this problem;
- We describe preliminary experiments that have been conducted in order to demonstrate the feasibility of streaming SQL optimization.

The remainder of this paper is structured as follows. First, we state the problem as illustrated by a running example (Sect. 2). Second, we list the challenges in adaptive optimization of streaming SQL queries (Sect. 3). Then, we present the preliminary experiments we executed on our running example to justify the potential benefits of the proposed approach to query optimization and discuss the results (Sect. 4). Finally, we discuss related work, including efforts on database and streaming query optimization (Sect. 5).

## 2   Problem Statement

This section illustrates the problem of streaming SQL query optimization using a running example of a query for a streaming system.

We are using the NEXMark benchmark [31] for our query. The NEXMark benchmark suite, designed for queries over continuous data streams, is an extension of the XMark benchmark [27] adopted for use with streaming data. The NEXMark scenario simulates an online auction system with three kinds of entities: people selling items or bidding on items, items submitted for auction, and bids on items. These kinds of entities will be referred to as Person, Auction, and Bid respectively. The original NEXMark benchmark includes eight queries that utilize the full spectrum of SQL features, but none of them contain more than one join operator. Unfortunately, the system used in our experiments can optimize the order of joins only. We extended the benchmark with the following query based on the NEXMark model:

```
1   SELECT P.name, P.city, P.state,
2           B.price, A.itemName
3     FROM Person P
4       INNER JOIN Bid B
5         ON B.bidder = P.id
6       INNER JOIN Auction A
7         ON A.seller = P.id
```

This query selects all the people who have joined the auction as both bidders and sellers. For each such person, their name, city, and state of residence are selected, as well as the price of each of their bids and the name of each item they are selling at the auction.

This query contains two join operators, which means that there are at least two ways to execute this query.

One of the logical plans (with substituted variable names omitted; using Apache Beam transforms as operators) for our example query is as follows:

```
1  LogicalProject(name, city, state,
2                        price, itemName)
3    LogicalJoin(condition, joinType=inner)
4      LogicalJoin(condition, joinType=inner)
5        BeamIOSourceRel(table=Person)
6        BeamIOSourceRel(table=Bid)
7        BeamIOSourceRel(table=Auction)
```

A physical plan derived from this logical plan produced for Flink executor is as follows:

```
1  BeamCalcRel(name, city, state, price, itemName)
2    BeamCoGBKJoinRel(condition, joinType=inner)
3      BeamIOSourceRel(table=Bid)
4      BeamCoGBKJoinRel(condition, joinType=inner)
5        BeamIOSourceRel(table=Person)
6        BeamIOSourceRel(table=Auction)
```

An alternative is the following physical plan for this query, with the two join operators in a different order:

```
1  BeamCalcRel(name, city, state, price, itemName)
2    BeamCoGBKJoinRel(condition, joinType=inner)
3      BeamIOSourceRel(table=Auction)
4      BeamCoGBKJoinRel(condition, joinType=inner)
5        BeamIOSourceRel(table=Person)
6        BeamIOSourceRel(table=Bid)
```

In unbounded data streams, elements can be grouped into windows based on event time or the number of tuples in each window, and each window can be processed similarly to a SQL table. Cost-based optimization requires statistical knowledge about the data, such as the cardinality of each window, which can be inferred from the element arrival rate in the case of streaming data. In our example, the first plan, where Person and Bid are joined first, and then the result is joined with Auction, would be preferable if the arrival rate of auctions significantly exceeded the arrival rate of bids, meaning that while many items have been getting put up for Auction, not many sellers have been making bids. If, however, after some time, sellers started making many bids, the second execution plan would have a lesser cost value. Thus, as data statistics change for the query execution, a previously optimal plan might become inefficient.

Due to the imprecision of data statistics, it is well-known that a plan optimal in terms of cost function is not necessarily optimal in actual resource consumption. To address this issue, several techniques known as adaptive query processing were developed [8]. Under these techniques, the execution is paused, and the query is re-optimized with more precise statistics, and then execution is resumed with the new plan. As the repeated optimization consumes a certain amount of resources itself, adaptive optimization makes sense for relatively long-running queries and is hardly applicable in data streams.

We use a different kind of adaptivity in our approach: the statistics collected during previous query executions are used to re-optimize the query for subsequent executions. As soon as the new plan changes, the query execution on subsequent windows is switched to the new plan.

This, as well as other specifics of SPEs, presents particular challenges in implementing global optimization of streaming queries, which we explore in the following section.

## 3    Challenges

In this section, we present the challenges in adapting SQL optimization techniques to data streams. We categorize each challenge as either research or engineering. The former category includes challenges that require further work, and we are not sure of the outcome. In contrast, the latter includes everything related to the incorporation of our approach into a production system.

### 3.1    Fetching and Predicting Data Statistics

Cost-based optimization requires statistical information on data in order to calculate cost function values for each plan. However, upon starting a streaming query execution, no information about the data, such as its arrival rate, is available. Therefore, to properly apply cost-based optimization to streaming SQL queries, it is necessary to collect data statistics over the course of query execution. Moreover, since we possess no definitive knowledge about the arriving data, we need to predict statistics for each next window based on statistics for previous windows to utilize an optimal plan for the upcoming windows. To this end, we identify two challenges in using statistics for streaming query optimization:

- **Engineering:** Statistical information on stream elements, such as their arrival rate, needs to be collected during execution at runtime without seriously affecting the performance of a distributed SPE.
- **Research:** We need techniques to predict statistics for upcoming windows based on statistics collected for previous windows. We expect that previous window statistics would present a decent baseline. However, this assumption requires further investigation.

Popular SPEs and frameworks for defining streaming workflows do not offer any statistics fetching or predicting. For example, the Apache Beam framework passes constant values to the query planner instead of any actual data statistics to Apache Calcite, a dynamic data management framework that implements its SQL processing functionality.

## 3.2  Using Statistics for Streaming Query Optimization

The API of the current state-of-the-art systems typically utilizes a consecutive approach to building a graph for query execution. First, a planner builds a logical graph. Then another component converts it to a physical graph, makes local optimizations, and finally passes it to an executor. Such separation leaves no opportunity to pass any data from the physical level to the logical level and adapt the graph to the new data, which makes any runtime adjustments to execution plans impossible.

The progress in applying various local optimizations to the execution graphs at the physical level (see [12] or Google Cloud Dataflow optimizer), there are no significant results in the logical level optimization yet, and the logical level allows to use of more complex optimizations than the physical level. Moreover, in this paper, we discuss distributed stream processing engines, which require additional consideration for query execution. To this end, we identify the following challenges:

- **Engineering:** It is necessary to adapt the API of current SPEs to pass statistics collected or predicted during the query execution at the physical level to the query planner at the logical level.
- **Research:** The relational algebra and the planner cost model require extension with new operators specific to distributed systems.

For example, a join operation should broadcast a stream with a low arrival rate of new elements to all the nodes in the system. In contrast, a high arrival rate suggests distributing different keys across different nodes. Therefore, the algebra should include a new *distribution* operator, and the cost model should include the estimate of its cost. The cost model should consider the latency due to communication between nodes as well. Such cost models have been well-researched in distributed databases [18] but not in distributed streaming systems.

## 3.3  Execution Graph Migration in Runtime

Streaming data is ever-changing: new data continues to arrive indefinitely, and the statistics for the following window elements might differ significantly from the statistics for previous windows.

Even if the optimal query execution plan was selected based upon statistics accumulated for previous windows, the new data statistics might be different enough to render the previous plan no longer optimal. Therefore, to adapt to the changes in data, it is necessary to identify the moment in time in which the previous plan is no longer optimal for the current or upcoming data and to migrate the execution to a new graph.

We identify the following challenges in migrating the execution graph in runtime:

- **Research:** Identifying the point in time at which it is feasible and beneficial to perform the graph migration is an open problem. First of all, it is necessary to estimate the costs of graph migration at the current point in time. Secondly, we need to establish what qualifies as substantial enough change in statistics to warrant graph migration.
- **Engineering:** The mechanics of graph migration, particularly for stateful operations such as joins and aggregations, in runtime have been researched [33]. However, most current popular SPEs do not provide such functionality.

One such strategy is the parallel track strategy described in [33]. The second graph can start the query execution and the first one, while the first one stops execution when the current window terminates. So all the subsequent windows are only processed by the new execution graph.

## 4    Preliminary Experiments

In this section, we describe the preliminary experiments that we conducted in order to demonstrate how the choice of a query plan affects the performance. We aim to show that a well-timed switch from an execution graph that is no longer optimal for the current data to a more optimal one would improve performance significantly. First, we present the experiment setup and the configuration used; then, we demonstrate our results.

### 4.1    Setup

For our experiments, we used the same query that was described in Sect. 2 and the Apache Beam implementation of the NEXMark benchmark model. In this implementation, each entity (`Person`, `Auction`, or `Bid`) is represented via a subclass of the `Event` class. Each event is generated by an unbounded source in accordance to the provided configuration, which includes parameters such as the arrival rate for each event, the $|Person| : |Auction| : |Bid|$ ratio, the time-based window size, etc.

First, we execute this query using the plan in which `Auction` and `Person` are joined first, and the result is joined with `Bid`; then, we use the plan in which `Person` and `Bid` are joined first (see Sect. 2). For each run, we use a different $|Person| : |Auction| : |Bid|$ ratio.

To evaluate performance, we measure latency and throughput for each window. For a join result, we consider *latency* to be the difference between the maximum arrival time of each of the rows making up the join result and the output time of the resulting row; then, we select the maximum out of the latency values of all the rows in each window. The throughput that we measure is *sustainable throughput*, i.e., the maximum events arrival rate that a streaming system can handle without the continuous buildup of latency.

We have conducted our experiments on a local machine equipped with a 1.4 GHz Intel Core i5-8257U CPU (4 cores) and 8 GB of memory using the Apache Flink runner.

## 4.2   Results

In the subsequent text the plan which joins `Auction` and `Person` first is referred to as *Plan 1*; the plan which joins `Person` and `Bid` first is *Plan 2*.

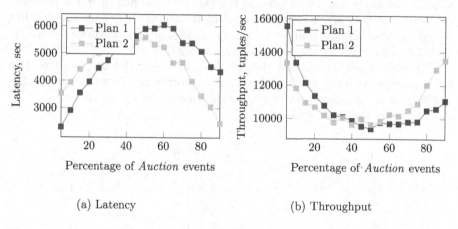

(a) Latency                                    (b) Throughput

**Fig. 1.** Latency and throughput for different ratios: out of 100 events, $|Person| = 5$, $|Auction|$ is the value on the $x$-axis, $|Bid| = 100 - |Person| - |Auction|$

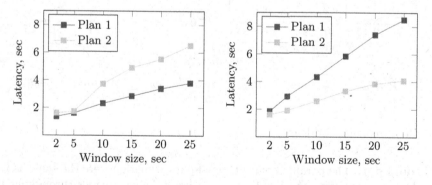

(a) $|Person|$ : $|Auction|$ : $|Bid|$ = 5:5:90

(b) $|Person|$ : $|Auction|$ : $|Bid|$ = 5:90:5

**Fig. 2.** Latency for different window sizes and $|Person|$ : $|Auction|$ : $|Bid|$ ratios

**Latency.** We generated 1000000 events with the arrival rate of 10000 events per second and time-based windows of varying sizes.

Figure 1a demonstrates how latency changes depending on data characteristics. The $|Person| : |Auction| : |Bid|$ ratio impacts arrival rate for each kind of entities, thereby influencing latency. As expected, the plan in which Person and Auction are joined first delivers better results when the arrival rate of Bid records significantly overwhelms the rates of Person and Auction (the 5:5:90 ratio is an example of such a case), while the plan in which Person and Bid are joined first works best for cases where the rate of Auction records far exceeds those of Person and Bid.

As Figs. 2a and 2b demonstrate, the latency does not grow linearly as the window size increases. This is due to the fact that the join operator processes the records as they arrive instead of starting to process them only after the last record in the window has arrived, thus the results are ready to emerge shortly thereafter the arrival of the last record in the window.

Figure 4a demonstrates how the difference in latency for the two execution plans changes with the window size. Since the difference grows as the window size increases, statistics-based optimization should provide an even bigger performance gain for larger windows.

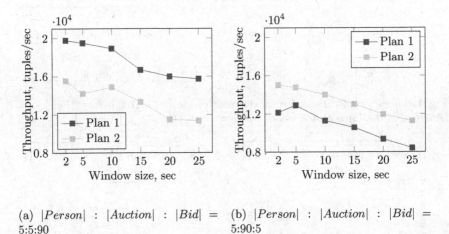

(a) $|Person| : |Auction| : |Bid| =$ 5:5:90

(b) $|Person| : |Auction| : |Bid| =$ 5:90:5

**Fig. 3.** Throughput for different window sizes and $|Person| : |Auction| : |Bid|$ ratios

**Throughput.** The parameters for throughput estimation were the same as for the latency estimation. As Fig. 1b shows, Plan 2 delivers higher throughput in case of the arrival rate of Auction significantly exceeding that of Person and Bid, while Plan 2 performs better in case of the arrival rate of Bid being significantly higher. This corresponds with the latency measurements. Throughput decreases with the increase of window size, as shown in Figs. 3a and 3b. Throughput difference remains constant and does not depend on the window size, Fig. 4b demonstrates.

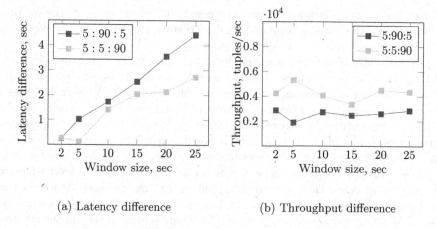

(a) Latency difference                    (b) Throughput difference

**Fig. 4.** Difference in latency and throughput for Plans 1 and 2

## 4.3   Discussion

The results of our experiments demonstrate that streaming query execution performance depends on the plan used for the execution, and the optimality of the plan depends on the data characteristics, which proves the necessity of adaptive optimization of streaming queries. Particularly, the first steps towards adaptive optimization should be predicting statistics for each window and performing runtime graph migration, since the results of the experiments show that even the current planners, such as the Volcano query planner [11] used in Apache Calcite, which are unaware of whether the data comes from a stream or a table, could use those statistics to produce a better execution plan. These two challenges will be the focus of our future work. After that, the planner can be enhanced by introducing distributed streaming systems-specific operators and their costs.

## 5   Related Work

We structure the related work into three areas: database query optimization, optimization of streaming queries, and predicting data statistics for queries using machine learning techniques.

Early efforts in database query optimization include the System R optimizer [29], which introduced a dynamic programming algorithm for finding an optimal execution plan; the Starburst optimizer [13], which proposed new rule-based optimization techniques; and others: a survey chapter on query optimization in relational databases can be found in [23]. Our particular interest lies in the areas of query optimization in distributed database systems and adaptive query optimization. The former has been described in [18]; a detailed survey [8] is a good reference for the latter. Adaptive optimization of streaming SQL queries is different from database optimization: while in databases, the execution plan is modified during the execution of a query, in streaming systems, the plan should be changed after a window has been processed but before the next window processing has started; moreover, database data is static during the query execution while streaming data is continuously updated.

An overview of streaming query optimizations can be found in [15]: most of these optimizations can be classified as rule-based and are applied statically at compilation time, although dynamic versions are listed for several of these; we are interested in adaptive optimization at runtime. Various works explore adaptive optimization of streaming queries: [4] focuses on finding the optimal order of pipelined filter operators, with possible reordering at runtime; it uses the currently known data statistics while we are interested in predicting the next window statistics and using those. Other works focus on physical level adaptive optimizations: one such study can be found in [12]; we are interested in logical level optimization instead. Another approach for physical optimization is discussed in [20]. It is based on the intermediate representation (IR) of the streaming queries, which supports a wide range of physical optimizations: operator redundancy elimination, operator separation, and fusion [15]. However, in this method, operator reordering-based optimization is limited compared to declarative queries optimization, which is more robust due to the algebraic properties of the relational model.

Previous works on execution graph reconfiguration in runtime, including Eddies project [3] and StreaMon [5], primarily study centralized stream processing, but, as we mentioned in Sect. 3, many issues originate from the distributed environment. The approach for state migration in distributed environment introduced in [16] can be applied as a building block for the general graph reconfiguration mechanism.

Some studies (e.g. [19,22]) use machine learning to predict optimal execution plans, while others explore join cardinality prediction. Database cardinality prediction techniques can be categorized as either query-driven or data-driven. Query-driven prediction models learn on sets of queries; among studies employing this approach are [7,17,21,24], which use neural networks. Data-driven prediction models, described in [14] and [32], are trained on data without queries and attempt to learn characteristics such as distribution of single attributes as well as joint distributions of multiple attributes. Neither of these approaches fit the streaming queries scenario: SPEs are executing the same query on each subsequent window, rendering query-driven approaches unusable, and the data is continuously updating, which means data-driven approaches are not employable either. Additionally, neural networks might not be the best choice of a learning model for a small amount of data contained in a single window; instead, it would potentially be more beneficial to use a model similar to [30], a fixed-size ensemble of heuristically replaced classifiers. Predicting statistics for streaming queries might yield better results than for database queries since the query is being run on different subsequent windows for an extended period of time, unlike in databases, where each new query is run separately from its predecessors. This could be advantageous in predicting statistics not only for the next input window but for intermediate execution results as well.

# 6  Conclusion

In this paper, we introduced a technique of adaptive optimization of streaming SQL queries in distributed stream processing engines. We highlighted the differences between adaptive optimization for database and stream queries. We argued the necessity of adaptive query optimization at both the logical and the physical levels. We identify the following challenges in the realization of optimization techniques for streaming queries: fetching and predicting data statistics, using predicted statistics for the query optimization, and migrating the execution graph in runtime once the statistics have changed enough that the previous execution plan becomes suboptimal.

We presented a running example of a streaming SQL query and performed experiments on this query and its different possible execution graphs to demonstrate the possibility of a gain in performance should the graph be adapted to the data statistics.

# References

1. Arasu, A., Babu, S., Widom, J.: The CQL continuous query language: semantic foundations and query execution. The VLDB J. **15**(2), 121–142 (2006)
2. Astrahan, M.M., et al.: System R: relational approach to database management. ACM Trans. Database Syst. (TODS) **1**(2), 97–137 (1976)
3. Avnur, R., Hellerstein, J.M.: Eddies: continuously adaptive query processing. SIGMOD Rec. **29**(2), 261–272, May 2000. https://doi.org/10.1145/335191.335420
4. Babu, S., Motwani, R., Munagala, K., Nishizawa, I., Widom, J.: Adaptive ordering of pipelined stream filters. In: Proceedings of the 2004 ACM SIGMOD International Conference on Management of Data, pp. 407–418 (2004)
5. Babu, S., Widom, J.: Streamon: an adaptive engine for stream query processing. In: Proceedings of the 2004 ACM SIGMOD International Conference on Management of Data, pp. 931–932. SIGMOD 2004, Association for Computing Machinery, New York (2004). https://doi.org/10.1145/1007568.1007702
6. Begoli, E., Akidau, T., Hueske, F., Hyde, J., Knight, K., Knowles, K.: One SQL to rule them all - an efficient and syntactically idiomatic approach to management of streams and tables. In: Proceedings of the 2019 International Conference on Management of Data, pp. 1757–1772. SIGMOD 2019, ACM, New York (2019). https://doi.org/10.1145/3299869.3314040, http://doi.acm.org/10.1145/3299869.3314040
7. Chen, L., Huang, H., Chen, D.: Join cardinality estimation by combining operator-level deep neural networks. Inf. Sci. **546**, 1047–1062 (2021). https://doi.org/10.1016/j.ins.2020.09.065, https://www.sciencedirect.com/science/article/pii/S0020025520309750
8. Deshpande, A., Ives, Z., Raman, V.: Adaptive query processing. found. Trends Databases **1**(1), 1–140 (2007)
9. Gedik, B., Andrade, H., Wu, K.L.: A code generation approach to optimizing high-performance distributed data stream processing. In: Proceedings of the 18th ACM Conference on Information and Knowledge Management, pp. 847–856 (2009)
10. Gedik, B., Andrade, H., Wu, K.L., Yu, P.S., Doo, M.: Spade: the system s declarative stream processing engine. In: Proceedings of the 2008 ACM SIGMOD International Conference on Management of Data, pp. 1123–1134 (2008)

11. Graefe, G., McKenna, W.J.: The volcano optimizer generator: extensibility and efficient search. In: Proceedings of IEEE 9th International Conference on Data Engineering, pp. 209–218. IEEE (1993)
12. Grulich, P.M., et al.: Efficient stream processing through adaptive query compilation. In: Proceedings of the 2020 ACM SIGMOD International Conference on Management of Data, SIGMOD 2020, pp. 2487–2503. Association for Computing Machinery, New York (2020)
13. Haas, L.M., Freytag, J.C., Lohman, G.M., Pirahesh, H.: Extensible query processing in starburst. In: Proceedings of the 1989 ACM SIGMOD International Conference on Management of Data, pp. 377–388 (1989)
14. Hilprecht, B., et al.: Learn from data, not from queries! Proc. VLDB Endow. **13**(7), 992–1005 (2020). https://doi.org/10.14778/3384345.3384349, https://doi.org/10.14778/3384345.3384349
15. Hirzel, M., Soulé, R., Schneider, S., Gedik, B., Grimm, R.: A catalog of stream processing optimizations. ACM Comput. Surv. **46**(4) (2014). https://doi.org/10.1145/2528412
16. Hoffmann, M., Lattuada, A., McSherry, F.: Megaphone: latency-conscious state migration for distributed streaming dataflows. Proc. VLDB Endow. **12**(9), 1002–1015 (2019). https://doi.org/10.14778/3329772.3329777
17. Kipf, A., Kipf, T., Radke, B., Leis, V., Boncz, P., Kemper, A.: Learned cardinalities: estimating correlated joins with deep learning. arXiv preprint arXiv:1809.00677 (2018)
18. Kossmann, D.: The state of the art in distributed query processing. ACM Comput. Surv. **32**(4), 422–469 (2000). https://doi.org/10.1145/371578.371598
19. Krishnan, S., Yang, Z., Goldberg, K., Hellerstein, J.M., Stoica, I.: Learning to optimize join queries with deep reinforcement learning. CoRR abs/1808.03196 (2018), http://arxiv.org/abs/1808.03196
20. Kroll, L., Segeljakt, K., Carbone, P., Schulte, C., Haridi, S.: Arc: an IR for batch and stream programming. In: Proceedings of the 17th ACM SIGPLAN International Symposium on Database Programming Languages, pp. 53–58 (2019)
21. Liu, H., Xu, M., Yu, Z., Corvinelli, V., Zuzarte, C.: Cardinality estimation using neural networks. In: Proceedings of the 25th Annual International Conference on Computer Science and Software Engineering. CASCON 2015, pp. 53–59. IBM Corp. (2015)
22. Marcus, R., et al.: A learned query optimizer. Proc. VLDB Endow. **12**(11), 1705–1718 (2019). https://doi.org/10.14778/3342263.3342644
23. Neumann, T.: Query optimization (in relational databases). In: Liu, L., Ozsu, M.T. (eds.) Encyclopedia of Database Systems, pp. 3009–3015. Springer, New York (2018). https://doi.org/10.1007/978-1-4614-8265-9_293
24. Ortiz, J., Balazinska, M., Gehrke, J., Keerthi, S.S.: An empirical analysis of deep learning for cardinality estimation. arXiv preprint arXiv:1905.06425 (2019)
25. Pitoura, E.: Query processing. In: Liu, L., Ozsu, M.T. (eds.) Encyclopedia of Database Systems, pp. 3026–3027. Springer, Boston (2018). https://doi.org/10.1007/978-1-4614-8265-9_860
26. Pitoura, E.: Query rewriting. In: Liu, L.,Ozsu, M.T. (eds.) Encyclopedia of Database Systems, pp. 3060–3060. Springer, New York (2018). https://doi.org/10.1007/978-1-4614-8265-9_863
27. Schmidt, A., et al.: A benchmark for xml data management. In: VLDB 2002: Proceedings of the 28th International Conference on Very Large Databases, pp. 974–985. Elsevier (2002)

28. Schneider, S., Hirzel, M., Gedik, B., Wu, K.L.: Auto-parallelizing stateful distributed streaming applications. In: Proceedings of the 21st international conference on Parallel Architectures and Compilation Techniques, pp. 53–64 (2012)

29. Selinger, P.G., Astrahan, M.M., Chamberlin, D.D., Lorie, R.A., Price, T.G.: Access path selection in a relational database management system. In: Proceedings of the 1979 ACM SIGMOD International Conference on Management of Data, pp. 23–34. SIGMOD 1979, Association for Computing Machinery, New York (1979). https://doi.org/10.1145/582095.582099, https://doi.org/10.1145/582095.582099

30. Street, W.N., Kim, Y.: A streaming ensemble algorithm (SEA) for large-scale classification. In: Lee, D., Schkolnick, M., Provost, F.J., Srikant, R. (eds.) Proceedings of the Seventh ACM SIGKDD International Conference on Knowledge Discovery and Data Mining, San Francisco, CA, USA, 26–29 August 2001, pp. 377–382. ACM (2001), http://portal.acm.org/citation.cfm?id=502512.502568

31. Tucker, P., Tufte, K., Papadimos, V., Maier, D.: Nexmark-a benchmark for queries over data streams (draft). Tech. rep., Technical report, OGI School of Science & Engineering at OHSU, Septembers (2008)

32. Yang, Z., et al.: Neurocard: one cardinality estimator for all tables. arXiv preprint arXiv:2006.08109 (2020)

33. Zhu, Y., Rundensteiner, E.A., Heineman, G.T.: Dynamic plan migration for continuous queries over data streams. In: Proceedings of the 2004 ACM SIGMOD International Conference on Management of Data, SIGMOD 2004, pp. 431–442. Association for Computing Machinery, New York (2004)

# Author Index

Printed in the United States
by Baker & Taylor Publisher Services